Green Up!

Sustainable Design Solutions for Healthier Work and Living Environments

Green Up!

Sustainable Design Solutions for Healthier Work and Living Environments

By
Stevie Famulari, Gds

Routledge
Taylor & Francis Group

A PRODUCTIVITY PRESS BOOK

First edition published in 2020
by Routledge/Productivity Press
52 Vanderbilt Avenue, 11th Floor New York, NY 10017
2 Park Square, Milton Park, Abingdon, Oxon OX14 4RN, UK

© 2020 by Stevie Famulari
Routledge/Productivity Press is an imprint of Taylor & Francis Group, an Informa business

No claim to original U.S. Government works

Printed in Canada on acid-free paper

International Standard Book Number-13: 978-0-367-27717-8 (Hardback)
International Standard Book Number-13: 978-0-367-27651-5 (Paperback)
International Standard Book Number-13: 978-0-429-29743-4 (eBook)

**Visit the Taylor & Francis Web site at
http://www.taylorandfrancis.com**

Green Up! *is dedicated to those before me who lived and researched green—for being ahead of your time.*

To those who are currently living and practicing green lifestyles on all levels—for choosing healthy and conscious choices.

And lastly, to the greeners of the future—for your innovation, ambition, and progressive creativity in green design.

Stevie Famulari, Gds

Contents

Contents

Acknowledgments

WITH GRATITUDE AND LOVE

Creating a book takes a team of people. Some of my fantastic team are mentioned here. To everyone that has helped, a big thank you beyond the words of this page. I am me, because you are you.

Alex—you made a challenging time much easier with such ease and grace. It made a world of difference, and still does.

AK—you are a fantastic friend and brilliant brainstormer. At times you have been the unknown voice in my head when I write. You are a gifted person who will always succeed in your own unique stunning way.

Bronx Charter School for International Cultures and the Arts—for giving me a space for inspiration in my community to live with green practices on a large scale. This school is a stunning example of pride, diversity, and success in the urban culture.

Charlie—for lunch every month for all these years and the chance to just sit, talk, laugh, and get to know you. Thank you for being you.

David—for being the best Gemini rifter who manifests his beautiful life, out loud, by the ocean.

My family—the many characters who have given me great moments to love over the years.

Herb—for giving me the time and support to write this book. And more, for always having my back. Forte al tavolo, ma ho i piedi piu duri.

Janna—what would I do without your voice and strength and drive?

Kimba—the southern belle, you remind me to feel that it is amazing to be a protagonist and that you can love me just as I am. Being your friend is always an adventure.

Lexi—for caring about the details and remembering to live your adventures.

Lisa—for every late-night dialogue and text that sends me love and light.

Missy—for light-filled years and beyond, with infinite love.

Nonni—for being a heroine who always has a spark in her eyes.

Paul and Shane—who have read every word from the beginning, and are a voice across the hallway or ocean of unending support and great meals.

Rizzo Group—for inspiring me with their office and people.

Shelly—who is with me when I write, and sends pure love, light, and laughter the entire time. You are a stunning soul that I am honored to know.

Author

Stevie Famulari, Gds teaches and designs public art, phytoremediation applications, storm water management, landscape architecture itstory, and specialty green courses. Her research in design explores the relationship, extension, and application of green designs to other professional fields. Stevie's PhD, ABD research is at RMIT. She received her master's degree in Landscape Architecture from SUNY College of Environmental Science and Forestry with a concentration in Fine Arts from Syracuse University. Her bachelor's degree of Fine Arts is from New York University.

Ms. Famulari's work focuses on greening designs and practices to create healthy spaces for living and working. By applying the science of phytoremediation to the art of landscape design, her works have aesthetic beauty as well as healing properties for both people and environment.

Stevie's projects include greenwalls; planted roofs; green remediation designs for interior and exterior applications; water remediation designed to benefit communities near oil fracking sites; designs for the Environmental Protection Agency in Colorado, North Dakota State University, and the Ghost Ranch Visitor Center for the Georgia O'Keefe Museum in Abiquiu, New Mexico; development of green design and policies on the UNM campus; flood control design for the Red River in North Dakota; and green design on numerous residential sites nationally.

Stevie Famulari's phytoremediation database of plants that clean the air, soil, and water of contaminants has been used by the EPA, landscape architecture, and engineering firms, and government agencies. She has been an investigator for grants which explore remediation design for oil drilling processes, the improvement of air quality, remediation design for communities, and interior green applications.

Her work in greening designs, practices, research, and education can be seen nationally and internationally through awards, lectures, presentations, and exhibitions at Harvard University Graduate School of Design, Plains Art Museum, San Diego Museum of Natural History, UC Berkeley, International Phytotechnology Society, MECA, UNM, UMN, ASLA, and AIA.

Stevie's work has appeared in hundreds of books, magazines, newspapers, and television programs, including Food Network Challenges and Specials, NBC Evening News, Oakland Tribune, World Entertainment News Network, The Oprah Winfrey Show, the Travel Channel, Good Morning America, CBS Early Morning News, *Smithsonian Magazine, Washington Post, The Post Standard,* Trust for Public Land, *Boston Herald, Berkeley Daily Planet, NY Village Voice, Santa Fe Reporter,* and *Star Tribune.*

Ms. Famulari has been a professor of Landscape Architecture and Urban Design for over a decade, as well as a director, a green artist, an author, and a researcher. She is currently a professor at Farmingdale State College, State University of New York, in the Department of Urban Horticulture and Design. She has also worked as a professor in the Department of Landscape Architecture at North Dakota State University.

Introduction

There are unique greening solutions and practices that you may use to help create a lifestyle shift. These practices will improve the health of your space and its occupant from a personal, business, environmental, and financial perspective. Short- and long-term impacts are important to keep in mind when moving towards healthy practices in lifestyles, choices, and site designs.

Green Up! addresses the various greening practices that can be applied to structures in urban and rural cultures. From the loft, to the neighborhood, the office space to the healthcare facility, this book outlines ways business owners, community members, designers, and residents can integrate scale appropriate green solutions into their lifestyles.

Green Up! Sustainable Design Solutions for Healthier Work and Living Environments includes illustrations and photographs to assist in understanding design opportunities for your space. I reside in the Bronx, New York, am a green designer and artist, principal of Engaging Green, and professor in Urban Horticulture and Design at Farmingdale State College, the State University of New York. My green designs are featured nationwide and my lectures and workshops have been held at Harvard Graduate School of Design, Smithsonian Museum, University of New Mexico, University of California Berkeley, and many other universities, businesses, and community groups.

Designer and city planner Ashley Kaisershot conducted an interview with me, and her interesting questions and my responses are featured throughout the book. Ashley Kaisershot, ASLA, has worked in architecture and engineering firms and has been a guest critic at North Dakota State University. She lives in Little Falls, Minnesota and works with communities throughout the Midwest as a planner. Ashley's perspective brings unique insights to inspire you to develop your own green understanding and design plan.

There is some terminology used in the book which is specific to the design field. Below are some terms explained in further detail:

Aeroponic plants

Aeroponic plants can be placed on a structure, such as a wooden log, and live without sitting in soil or water. The most common examples are plants that are sitting on logs in restrooms. These plants still need to be misted with water and nutrients; however, their roots do not need to sit in a soil or water base to grow.

Daylight bulbs

Light is measured in degrees Kelvin. Bulbs that are less than 5,600–6,000 degrees Kelvin may be tinted yellow, green, blue, etc., based on the degrees. Pure white light, daylight, is beneficial because it reaches and uses the full range of rods and cones in your eyes. Pure white light puts less stress on the eyes, allows more accurate colors, and provides light which makes it easier to see the space.

Greenwalls

Greenwalls are walls in which plants are grown. These walls can either be interior or exterior. The walls typically have an irrigation system designed into the structure. An illustration of how a greenwall works is found in Chapter 2.

Greenwalls with hydroponics and soil

There are two basic ways plants in standard greenwalls are grown: in soil with soil pockets or planters; or hydroponically in a capillary pad (essentially a sponge) which is watered. There are some plants which cannot grow hydroponically and must be grown in soil to be effective. There are plants, however, that can grow in either in a soil or water base. Greenwalls can use both systems in a design.

Grey water

Grey water is the water which comes from sinks, showers, dishwashers, washing machines, or the overflow from coffee makers. Grey water can be used for outdoor or indoor irrigation of plants which are grown in soil, water, or aeroponically. Plants are naturally able to use water which may contain soap or other detergents.

Hydroponics

Plants grown hydroponically are plants which grow in water. The roots of the plants either sit in water or on a capillary pad which is kept moist through a watering system. Nutrients are added to the water for the benefits of the plants. Not all plants are able to be grown in a hydroponic system as some plants live best in soil.

Itstory

Itstory is a word proposed in recent times to be a collaborative narrative; it is a response to people choosing herstory rather than history (his-story). As history and herstory are exclusive to certain genders, itstory is inclusive of not only both genders, but also animals and plants. It creates a more collaborative narrative and is inclusive of greater diversity.

Phytoremediation

Phytoremediation is the use of plants to uptake contaminants from the soil, water, or air. The contaminants may be metals, radionuclides, biodiesels, organics, or other materials. Plants typically take in the contaminants through their roots, though some plants also take in some contaminants through their leaves or structure.

Permeable Pavers

Permeable pavers are a ground covering of concrete stones which interlock with spaces in between the stones. The spaces between the stones are planted with plant materials or hold smaller stones. These spaces allow water to filter through to underground aquifers. Permeable pavers come in a variety of patterns and can be used in parking areas, walkways, and other sites.

Illustrations of applied greening are located throughout the book to help inspire your own goals and design, and then transform them to reality. Images are credited to the author unless otherwise noted. Professionals are available to help you with understanding your site conditions and opportunities for green solutions. With a myriad of options to choose from, the best solution is the one that works for you and all the users of the space, for both short- and long-term growth. A change to a healthy green lifestyle can be part of any design and may grow and expand as your desire for green practices expands.

Green Up! breaks down the misconceptions of the complexity of sustainability and green practices. This book provides illustrations and size appropriate green solutions you can incorporate into your lifestyle. Greening is a lifestyle change and this instruction guide lets you know how easy it is to transition to the green side. Designing green is about creating your site as a space for inspiration, safety, health, love, and growth.

Chapter 1

Apartments, Lofts, Studios—Interior Residential

INTRODUCTION

Green Solutions for Interior Residential Greening

There is great opportunity for fun and personal style with interior green design ideas in apartments and lofts. This can include vertical greenwalls, planted dividers, art pieces, grey water features, interior lawns, sculptures, and horizontal pieces on surfaces. Designs can be on the floor, above cabinets, and even on the ceiling. There are many unique spaces, sizes, shapes, forms, and materials to explore.

The designs are limited only by your imagination. There is always a solution to get plants and green practices into your space. Consider using materials that are local, recycled from other sites or site-specific.

Make the design yours. This chapter gives you places to start, understand, and build upon for your apartment and loft design.

Existing Site Conditions

Understanding the needs and wants of your site means understanding its scale, materials, and use. It also helps to understand the people who live in the apartment. Recognizing all these factors, along with sunlight direction, air leaks from windows, and other physical qualities of the space helps you to improve the space with a green healthy design. Understanding your site and air quality contaminants help as the first step to improve it with greening designs and practices, changes in lifestyle, or other healthy approaches.

Some site conditions and materials are not likely to be altered frequently. For example, the materials of existing walls and windows are not changed often. We also enjoy using our favorite inherited dining room table or other sentimental furniture where other materials such as curtains, couches, or wall paint colors change more frequently.

We completely understand that changing designs and upgrading is part of the enjoyment of living in or owning a space. When you are changing lights, fabrics, furniture, paints, fans, and other elements, consider objects that are sustainable such as those which use materials which are low in VOCs (volatile organic compounds) and other contaminants, or objects which use natural and sustainable materials.

Look at the date your apartment or building was built. Based on the year it was built, the materials used in construction can help determine existing contaminants. Examine your apartment—what are the objects made of: the walls, the furniture, and the surrounding neighborhood conditions. Household cleaners, glues, beauty sprays, paints, and sealants used to restore old furniture can also be changed to natural choices or to resources that have a low chemical content. When choosing furniture and cabinets, the woods, metals, and sealants are all areas where you can talk with a professional to choose the healthiest option.

WHAT TO DO NEXT

Questions to Ask about Your Site

Once you have looked at the objects in your home, look at the space itself: the walls, the shape, the architecture, air circulation, artificial light, and sunlight direction. In an interior residence, asking yourself questions to understand your own wants and needs for the site is important. The questions may be basic, such as, how much sunlight is there in my space and do I need to add daylight bulbs? The questions may be more complex, such as, what elements are in my design that affects air quality and what can I do in my design to improve it?

Then consider what you can change or improve. You may even begin to look for unique resources for designing your site, or a designer to help you bring your ideas to life. Think outside the box—in fact, shatter the box entirely. Look through this chapter and other chapters to help inspire you.

To help illustrate the application of green designs and working with site conditions further, two different style apartments are used to illustrate what to look for in a site and how to apply greening designs and practices to them. All sites have conditions which produce air contaminants that can be improved with green designs. These examples are to help you to understand and look at your site carefully with a unique green perspective. Ashley Kaisershot (AK) interviews the author Stevie Famulari (SF) throughout the chapter with questions to help further understand topics.

The two sample homes include an urban apartment in New York City and a two-story house converted into four units in Fargo, North Dakota.

Example A—Urban Apartment in the Bronx, NY

An urban building

- The building was built in the south Bronx in 1926/1927 as a co-op. The residents typically own and live in their apartments long term.
- Located in a metropolis, the conditions of urban sounds and urban air quality exist.
- Urban air typically contains a higher concentration of contaminants and PM (particulate matter) such as SO_2 (sulfur dioxide), NO_2 (nitrogen dioxide), lead, carbon monoxide, and ground level ozone.
- These contaminants are a result of vehicle engine fumes, fossil fuels, industrial production, and natural processes such as wind-blown dust.

Images 1.1, 1.2.

Bronx, New York, six-story, 149-unit apartment complex.

Images by Heidi Solander

An urban apartment

- The apartment is located in a six-story building, with 149 units.
- The building has two planted and concrete courtyards with three of the four sides of the building facing a street and/or courtyard.
- All of the six windows of this apartment face a wide alley between two buildings.

- On the fifth floor, the light is pleasant and sunny during the daytime.
- With a visible line to the neighbor's windows, shades are sometimes pulled to mask a direct view.
- The apartment is approximately 600 square feet, with an exposed brick wall and ceiling fan in the main room and bedroom.
- The apartment has been renovated in recent years, including new cabinets and appliances.
- There is a gas stove without a hood ventilation, no dishwasher, and wood floors.

The neighboring units

- In a neighboring apartment there is a person who smokes out the window. The cigarette smoke comes in through the windows in the sample apartment.
- It has been discovered that the household dust of smokers contains lead, pesticides, PAH (polycyclic aromatic hydrocarbons), allergens, nicotine, and more than 4,000 chemicals. Second-hand cigarette smoke also contains these contaminants.
- The walls, furniture, clothes, toys, and other objects in the space all absorb the contaminants within minutes of lighting a cigarette or cigar. The contaminants are re-emitted into the air for months as it reacts to vapors. The apartment acts as a reservoir for the ETS (environmental tobacco smoke) long after the cigarette has been burned.

The Green Design Can Improve

Knowing all these factors, the resulting design can improve air circulation with fans that are on routinely, as well as help reduce the contaminants from the gas stove, urban air, the neighbor's cigarettes, household cleaners, paints, furniture, cabinets, and wood sealants used on the floor. Additional design solutions can be used for the dish drying rack, and for composting and recycling.

Images 1.3, 1.4.

Bronx, New York, one bedroom apartment on the fifth story.

Images by Marc Newberger

Images 1.5, 1.6.

Bronx, New York, one bedroom apartment on the fifth story.

Images by Marc Newberger

*Example B—Two-Story House That Has Been Converted
into Multiple Apartments in Fargo, ND*

A converted house

- The residence was constructed to be a single-family dwelling in 1954 with concrete, wood, plaster, glass, metal, and shingles.
- In the 1970s it was converted into a quadruplex apartment building with mostly students from the nearby university renting the apartments.
- It is a two-story building that is 30.5 foot × 50 foot with a roof slope that is approximately 9:12.
- The plant materials found around the complex are turf grass, hollyhocks, three mature trees, and a small vegetable garden.

Images 1.7, 1.8.

Fargo, North Dakota converted house to apartments.

Images by @LisaEleni.com

A rural apartment

- The average turnover rate of the student renters is between 2–3 years.
- Each apartment is 600 square feet on average.
- There is a shared staircase to get to the apartments.
- The apartments have radiant heat throughout.
- There is a lack of air circulation creating dust throughout the apartments and shared stairwell.
- The windows leak air from outdoors. This affects the temperature inside the apartment. Notably, the winters in Fargo routinely get to temperatures below 0-degrees Fahrenheit.
- The gas stove (at the time of writing) emits a faint scent of gas. The landlord has not fixed it as of the date of writing. Gas stoves emit NO_2 (nitrogen dioxide), CO, (carbon monoxide), and HCHO (formaldehyde), each of which can exacerbate various respiratory and other health ailments.

The Green Design Can Improve

Conscious of all these factors, the resulting design can improve the air quality and air circulation with fans, as well as help reduce the contaminants from the gas stove, air, household cleaners, paints, furniture, cabinets, and wood sealants used on the floor. Additional design solutions can be used for the dish drying rack, and for composting and recycling. Additionally, there is a high rate of wear and tear from the high turnover of renters. The wear and tear can be reduced by having a permanent interior planted green design built directly into the space. This can raise the quality of the renters (with higher quality units and higher rental prices) and keep the units in better condition long term.

Images 1.9, 1.10, 1.11.

Apartment in a converted house in Fargo, North Dakota.

Image 1.9 by @LisaEleni.com

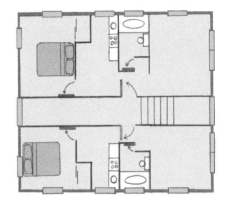

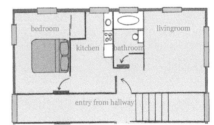

Images 1.12, 1.13, 1.14.

Bronx Apartment with greening applications: Main living area, dining island, and kitchen of the apartment.

Image 1.13 by Marc Newberger

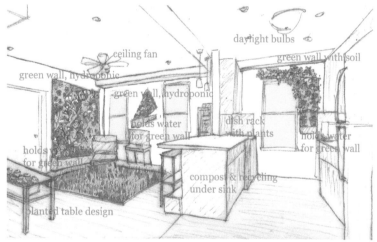

Image 1.15.

In the Bronx apartment, there is a hydroponic greenwall in the main area, one soil-based potted plant wall piece, a table with planters built in, and a small vertical arrangement of plants. The large greenwall is hydroponic, with no soil. Each of these design elements has their own water, pump, and lights set on a timer. This apartment contains its own grey water, which can be used for watering.

The lawn is in-ground planted grass in shallow soil, lined at the bottom with a waterproof liner. The watering system for the lawn is set on a timer. The grass and other plants add a fresh scent to the space. Clover and wild flower mix and can be added to the lawn as desired.

The nutrients needed for plant growth are put into the water system. These nutrients can be bought by the gallon and used on a schedule based on the amount of plants and water. Typically, a capful of nutrients is put in the water once a week. An exact determination of the quantity is based on the scale of the wall, the amount of water used, the amount of water which naturally evaporates, and which plants are in the design.

Plants can be replaced as needed (as in all greenwalls). The plants chosen throughout the green design in the Bronx apartment have been chosen for the improvement of the air quality as well as for aesthetic purposes as noted.

The plants used to address formaldehyde and carbon monoxide include,

Spider Plant (*Chlorophytum comosum*)
Chrysanthemum (*Chrysanthemum x morifolium*)
Orchid (*Dendrobium taurinum*)
Warneckei (*Dracaena deremensis 'Warneckei'*)
Red-Edged Dracaena (*Dracaena marginata*)
English Ivy (*Hedera helix*)
Lady Palm- (*Phapis excelsa*)
Snake Plant- (*Sansevieria trifasciata*)
Peace Lily- (*Spathiphyllum wallisii*)

AK: *Why is __air circulation__ important for a healthy space?*

SF: Creating an air circulation system is important for the air to be cleaned through plant materials and to circulate the clean air throughout the space. Radiant heat allows for no circulation, as is the case with the Fargo apartment. The radiant heat circulation and air quality can be improved with something as basic as ceiling fans throughout the apartments and stairwell.

The Bronx apartment has two radiators and heating pipes. Two separate air circulation plans are applied—a fan in the main space and a second ceiling fan in the bedroom. With both rooms having their own green designs, this circulation pattern is workable.

Radiators and heating pipes are located both in the main room and the bedroom. With the ceiling fans

on and the windows open, air circulation in this space is moving well. The windows are open most of the year. In winter the heat can be too warm, and it cannot be regulated or adjusted within the individual apartment—the heat is set for the building as a whole. With 149 units, some apartments are more effectively heated than others due to age of windows, directional wind, and distance and flow of the hot water pipes. The slightly open windows and ceiling fans circulate the heat in the winter and cool air in the nighttime in the summer.

The plants used to uptake nicotine include,
  Peppermint Plant (*Mentha piperita*)

Plants used to uptake PAH (Polycyclic Aromatic Hydrocarbon)

  Buffalo Grass (*Bouteloua dactyloides*)
  Canadian Wild Rye (*Elymus canadensis*)
  Red Fescue (*Festuca rubra*)
  Ryegrass (*Lolium multiflorum*)
  Empire Bird's Foot Trefoil (*Lotus corniculatus*)
  Alfalfa (*Medicago sativa*)
  Western Wheatgrass (*Pascopyrum smithii*)
  Reed Canary Grass (*Phalaris arundinacea*)
  White Clover (*Trifolium repens*)
  Tall Nettle (*Urtica procera*)

To uptake lead

  Indian Mustard (*Brassica juncea*)
  Canola (*Brassica napus*)

SECOND-HAND CIGARETTE SMOKE

Image 1.16.

The effects of the cigarette smoke from the neighboring apartment can be mitigated with the plants listed. Putting these plants near the window will help with long-term effects. For a short-term solution—as when the cigarette smoke is coming into the apartment for the five minutes in which it is burning—a design that includes a scented oil such as lavender and a fan near the window can help with the strongest scents.

This design utilizes a window fan with a scented screen for a soothing temporary effect. The two large greenwalls are easily moved on wheels, allowing the pieces to be placed in front of the windows as screens, or moved to the sides as wanted.

*AK: How do I find out what **air contaminants** may be on my site? Do I have to get my space tested?*

SF: Air contaminants come naturally as part of the decaying and transformative process. For example, as wood decays it releases some contaminants which changes the air quality around it. There are many variables and elements which do not produce harmful contaminants, whereas some are more harmful. An example of a transformative process is paint fumes. Fresh paint produces VOCs (volatile organic compounds). There are some plants which take in VOCs from paint without harming the plant and clean the air naturally.

Other examples in apartments include common sealants in furniture, fabrics in furniture, gas stoves, mold, cleaners used for kitchens, bathrooms, floors and carpets—all of which affect air quality. Beyond these, there are outdoor air elements which get inside through windows—such as car exhaust fumes, cigarette smoke, and outdoor urban air elements. All of these affect air quality. Professional companies can test your air for specific contaminants and levels. You can look around your space and begin to determine the materials, sealants, cleaners, and other elements of your site that are likely to contain chemicals. Objects in your space, their materials, and liquids all affect air quality. If it contains a chemical, it is affecting your air quality.

For both the Bronx and Fargo examples, a testing schedule can be implemented. There are devices that can be placed throughout the home which have computer apps that determine contaminants and provide air quality reports. There are also companies which can test for air quality. Resources for testing air quality can be found in local and regional companies. One suggested schedule approach is:

- Initial testing of the air prior to design installation.
- Testing 30 days after the design and plants are established.
- Testing every six months thereafter, or as needed if other site conditions change.

DISH DRYING RACK

Image 1.17.

This is a basic design for a wall mounted dish drying rack. The water drains from the area where the dishes are drying to the planter in the front. You can add more water as needed based on the plants you choose. Dependent on the specific plants, the planter can either have soil or be hydroponic with marbles or loose stones; the stones act as a structure for the roots to hold on to.

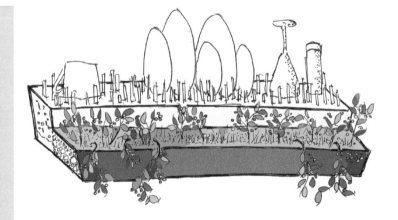

Plants, along with air circulation and choosing natural cleaners, can improve air quality and the health of the users in a space.

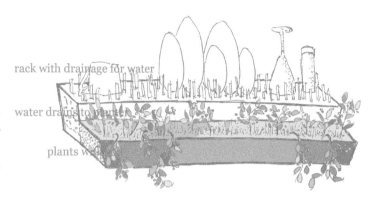

rack with drainage for water

water drains to planter

plants w

*AK: How do I choose the plants for my design? Is there a reason to choose some over others? Are **some plants more effective** for specific reasons?*

SF: It is a natural process for materials and objects to decay and change form. Some decay processes take a very short time, such as food decay. Other materials however, have a longer decay process, such as sealed wooden furniture. During part of the natural decay process, the contaminants become airborne and can reach a stage where they are harmful to breathe in. That time period may vary depending on the situation: fresh paint fumes can last a few days in this harmful state whereas sealants and chemicals can be airborne longer.

WALL MOUNTED DISH DRYING RACK

Image 1.18.

Even your <u>dish rack</u> could be a planter. A wall mounted dish rack with an attached planter can utilize the water from the drying dishes and divert it into the planter. For those people that do not have a dishwasher, consider designing your own dish rack planter to utilize the collected water for supplemental water for the planter. Even if the water is not fully fresh and has residual dish soap, plants are able to absorb this water easily. This can be a fun conversation starter with your guests.

Plants, however, can take in contaminants from the air, water, and soil and not be harmed, while continuing to grow. This process of using plants can improve the air, water, or soil quality naturally. Using plants to clean the air, soil, or water is known as phytoremediation. You can use plants which remediate some of the specific needs of your space, as well as those that work for overall cleaning benefits.

You may choose to enhance your site by not adding unnecessary chemicals to your space with thoughtful choices in cleaning agents and in the materials and sealants used in your furniture. If your design materials happen to add more contaminants, consider creating a design that tips the balance toward more plants and green practices across the overall site space.

When exploring design options, consider plants which may be suitable for specific contaminants are in your site. There is a phytoremediation database that you may use as a resource which can be utilized to find plants which uptake contaminants on sites. The database of contaminants and plants for phytoremediation is a way to aid cleaning up by greening up. The database can be found at www.engaginggreen.com through the phytoremediation database link. The database is updated to show more studies as they arise.

There are additional reasons to choose specific plants, aside from air quality improvement, sunlight, or other site conditions. Plants are also chosen for scent, visual interest, colors, form, personal memories or sentiment, spirituality, and cultural reasons. These reasons are just as important as functional reasons such as air quality. Both form and function are applicable in your design. Choose plants that make you feel good. Plants change the energy of a space, bring calmness, and exemplify interesting daily changes.

AK: Is there a reason to **aerate the water** (add air to the water)? How do I use an aerator?

SF: Aerating water helps to add air to the water, circulating the water even if it is in a container used for the greenwalls. This is helpful because connecting air to water helps to reduce or remove dissolved gases and oxidizes metals and VOCs. This is beneficial for plants, enabling them to get more nutrients from the water. There are aerators you can buy, such as those used in the fish tanks. Choose the one that is suitable to the size of your system.

APARTMENT COMPOSTERS

There are multiple choices for purchasing electric and non-electric interior composting units. Based on your needs, such as the amount of food scraps from your household, as well as the space available, you can choose to purchase a pre-fabricated design or create your own.

Interior composting is safe, easy, and can provide rich nutrients for your plants in the soil. Creating your own design does not have to be complicated and can be unique and specific for your needs. Some apartment complexes, such as the one in the south Bronx, have a shared composting unit for the building.

Images 1.19, 1.20.

Fargo converted house to apartments—green apartment design;
main living area of the apartment

GREENWALLS AND PLANTERS FOR PHYTOREMEDIATION

In the Fargo apartment complex an interior greenwall is incorporated into the stairwell to clean air contaminants through phytoremediation. Within each of the four apartment units, planters are incorporated with a grey water system and plants are effectively used for phytoremediation of the air.

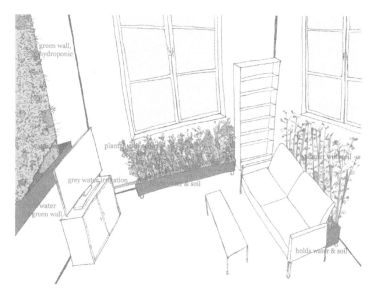

The plants for this design may include the same ones noted for the Bronx apartment. Additional plants include:

<u>Plants to uptake gas stove contaminants</u>

Bamboo Palm
(*Chamaedorea seifrizii*)
Red Fescue (*Festuca rubra*)
Tall Fescue (*Festuca arundinacea*)
Soft Rush (*Juncus effuses*)
American Vetch (*Vicia Americana*)

Image 1.21.

The planters have a pipe for watering, linking them all together within one room. This allows for water overflow from one planter to another, while still allowing for light and irrigation timers to be individual for each planter.

A pipe system is designed to follow a series of pipes within each apartment to carry grey water from the kitchen and bathroom sinks and bathtub to the planters for watering. With five people in the building, each apartment produces approximately 53 gallons of grey water daily, most of which is used for the green design in the apartments and the shared stairwell design. Planters can retain their own water from the grey water sources. When the planters need more water, a person can either set a timer to release more grey water from the source or water can be poured directly into the water tank of the design.

The apartment planting system uses all of its own grey water from each apartment unit. There is also a backup area for the water should it ever need to be quickly drained. You can drain water down the tub if needed. To contain and collect this water each day, there is a water storage basin beneath the planter equipped with a pump, filter, and watering heads to water the plants on a timer every day. The planters can be movable, so you can adjust the design layout if needed.

AK: What if you have **<u>windows that leak</u>** *or have drafts that are noticeable? Do leaks affect the temperature of the space which also affects the plants?*

SF: If your windows leak and affect the temperature of the space in the extremes of summer and winter, sealing the windows will help for many reasons. It will help with the temperature of your space as well as leaks from outdoor air contaminants. This is important if you live in an area where car exhaust fumes, or unwanted cigarette smoke from a neighbor may affect you. You can still create designs even with window leaks, choosing plants that are suitable for the conditions.

AK: *What is **grey water** and is it safe to use in plants?*

SF: Yes, grey water is safe for plants. Grey water is the water that comes from your sinks, tubs or showers. Black water is from toilets and is not reusable. Grey water may have soap or diluted cleaning agents in it but plants can use the grey water easily in both soil-based systems and hydroponic systems. You can supplement the grey water with fresh water.

The grey water sources in each apartment can be used in the planted designs by using excess water from the kitchen and bathroom sinks, bath, and shower, and creating a series of tubes to planters. The tubes or piping can be designed functionally and aesthetically inside the home. If grey water cannot be piped across an apartment, it can be stored under the sink and used when needed for watering the plants. This can be done manually if the apartment does not allow for tubes to be run through the space.

AK: *I have some expired **medicines**, aspirin, and liquids, in my home. Can I put those in the greenwall water rather than throwing them out or dumping them down the sink?*

SF: Yes, you can put the medicine in the soil or hydroponic design. Over time the plants, water, and soil will naturally break down the medicine and pass it through the plants. The benefit of using plants for medicine disposal is that the medicines do not get added to the garbage. Similar to composting, this choice can decrease the amount of garbage produced by a household.

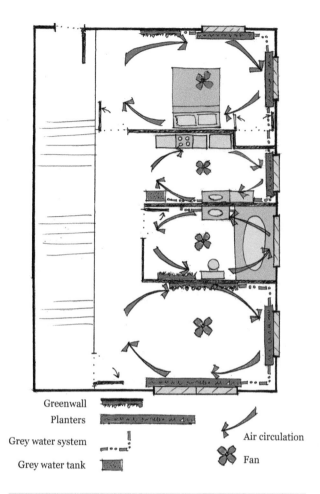

Greenwall

Planters

Grey water system

Grey water tank

Air circulation

Fan

AIR CIRCULATION DESIGN

Image 1.22.

In the Fargo design, circulation of air to the stairwell is integrated into the design. This is needed as there is little air circulation with radiant heat, as well as there being no ventilation system. Experiment to make sure the air moves in patterns that are felt throughout the individual apartments as well as the common area of the stairwell. This will help the air circulate through the plant materials for improved air quality.

In the Fargo design, ceiling fans in the individual rooms create circulation in each area, with the windows assisting during months when they may be open. The common stairwell has a window at one end and a door at the other allowing for circulation with fans.

Images 1.23, 1.24.

Fargo, house converted into apartments—green stairwell design

Shared stairwell for access to second-floor apartments

Green Stairwell Design

There are multiple individual planted greenwalls in the stairwell. They are in rows running along the walls, with the length of each row based on the space. Each of the planted rows have their own pump, light source, and water system set on a timer. Low-light plants are best suited for this site.

green wall design in stairwell

irrigation with grey water

shelves with plants

The plants may include: for low light

Snake Plant (*Sansevieria trifasciata laurentii*)
ZZ Plant (*Zamiocalcus*)
English Ivy (*Hedera helix*)
Pothos (*Epipremnum aureum*)
Leopard Lily (*Dieffenbachia*)
Philodendron (*Philodendron*)
Rex Begonia (*Begonia rex*)
Chinese evergreen (*Agloaonema*)
Peace Lily (*Spathiphyllum*)

Images 1.25, 1.26.

A loft design—green designs act as artworks as well as vertical and horizontal gardens.

A large loft with open space, large windows, and hours of sun.

There are both hydroponic and soil-based greenwalls in this large space. The pieces are both gardens and art-works. The planted table includes herbs and edible plants.

Each piece has its own timer, lights, pump, and water system. Behind the wall pieces is a liner to protect the wall from moisture. A combination of grey and fresh water can be used for this design. The plants are kept moist but not soaking in water through the timed system.

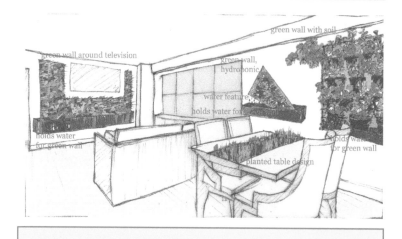

Plants in this design include:

For colorful blooming aesthetics: Grass Seed Mix, Wild Flower Mix, Ivy, Ixora, Streptocarpus, Roses (*Rosa*), Winter Wheat (*Triticum aestivum*), Calla Lily (*Zantedeschia aethiopica*), Tulips (*Tulipa*), Flowering Maple (*Abutilon hybridum*), Oxalis (*Oxalis triangularis*), Jasmine (*Jasminum*), Shrimp Plant (*Justicia brandegeeana*),

Herbs include: Basil, Mint, Parsley, Rosemary, Sage, Thyme

HERB TABLE AND GREENWALL WATER FEATURE

Image 1.27.

The table has a shallow center panel with herbs. The table has soil with its own irrigation system, which could be used for growing plants with enjoyable scents or for growing your favorite herbs.

The triangle-shaped wall design has a space where water can be heard falling into the reservoir. You can design a gentle sound or more dramatic sound based on the distance the water has to fall. The longer the distance and the longer the surface, the more dramatic the sound.

AK: *Does the direction of the window and __sunlight__ matter? Can I change the light in the space?*

SF: The amount of sunlight hours and the direction of the sunlight affects the plants you choose. However, you can supplement this by adding your own artificial light. Daylight bulbs are resources that can be used to supplement your existing natural daylight—even improving the light, as daylight bulbs are full spectrum. Full spectrum or daylight bulbs mean the light is pure white light—not yellow or green tinted. Companies such as Ott technologies and other sources sell daylight bulbs. Daylight bulbs are 5,600–6,000 degrees Kelvin. More information about the positive effects of daylight bulbs can be found throughout the book.

AK: *How do I __maintain the design__? How do I know what to set the timer to for lights and water?*

SF: Decisions such as the timer settings for light and water are based on the scale of the design, plant choices, and where you live. The weather, humidity, plants chosen, size of the design, and scale of the space are all elements to factor in to the timer. I lived in the desert region of Albuquerque, New Mexico, where it was very dry. The greenwalls needed to have the water run more frequently than those of similar scale in Fargo and New York City. This is because the air has different humidity levels, even indoors. In dry areas or dry seasons, you may need to adjust the timer to add more watering time and/or lights on the plants. This is part trial and error. In Albuquerque, New Mexico the timers were set to run water 60–65% of the day; whereas in Fargo, North Dakota, the timers were set to run water 50–55% of the day. The general rule of thumb is that the plants should stay moist but not soaking wet, nor should they dry out entirely. Setting the timer for this balance may mean equal times of having the water on/off, (one hour on, one hour off) or adjusting as needed, even seasonally.

Images 1.28, 1.29.

Albuquerque hydroponic greenwall

Main living and dining area with greenwall of grasses

ALBUQUERQUE GREENWALL PROCESS

This design, approximately 9 foot × 12 foot, is a site-specific fully hydroponic greenwall for a residence in Albuquerque, New Mexico. It was first planted with a patterned set of seeds in the greenwall. Some plants and seeds were changed over the years, including sewing in moss and ivy. Eventually, grasses, herbs, and wheat were chosen for long-term growth.

AK: *How do I know **how much weight** my wall can hold? How much does water weigh?*

SF: Each wall is different. Creating a design that is attached to a wall, free standing, or a combination of both is based on your site. Accounting for the weight of the water and soil is also important to consider. For example, one gallon of water weighs 8.36 pounds. Whereas, 1 cubic foot of water weighs 62.4 pounds (because it contains 7.48 gallons). Even if the water is running through the plants, the total weight of the design is important to consider, particularly in larger wall pieces. Seeking a professional may be best for complex designs.

AK: *Is there an advantage to **soil versus hydroponics for plants**?*

SF: They are both different mediums and produce different effects. Typically, flowering and showy plants grow easiest in soil because of the nutrients in the soil. In hydroponic designs, nutrients are added to the water. Once the best nutrients are found, it is easy to maintain. Without soil many of the bugs associated with soil are less likely to appear. Though bugs may happen in hydroponic designs, there are natural ways to remove the bugs. One trick is to use a small amount of citric acid (orange juice or lemon juice) in the water. The 9 foot × 12 foot greenwall needs only two tablespoons of orange juice to be effective. This technique can also be effective in plants in soil.

Choosing the most effective medium is based on your wants and needs. You can also create a combination of these mediums for a striking design.

Image 1.30, 1.31.

Most of the plants chosen for this wall are self-seeding- meaning they grow and re-seed themselves. Some plants on the wall are not self-seeding and are replanted after their life cycle. This allows for a change of some showier flowering plants to change the look of the design when wanted.

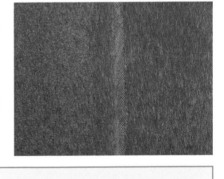

AK: *Where do I get **nutrients** for the hydroponic system?*

SF: You can find gallon mixes of nutrients in garden stores, hydroponic stores, or online. There are a variety of brands and nutrients that you can choose from. Talk with a specialist about your specific plants and needs.

The greenwall has a timer which controls the lights and water, set accordingly for the dry air of the desert region of Albuquerque, New Mexico. The water runs to keep the (blue) capillary pad which holds the plants moist, never drying out or soaking wet for a day. In the driest season, the submersed pump runs every 2–3 hours for 1 hour of water flow.

UNIQUE DESIGN FEATURES OF THIS CORNER GREENWALL

Image 1.32.

a. Adding tranquil water sounds in this greenwall design can be easy. One way is to cut the edge of the pad—where it meets the water—to create a gap where the water edge is lower than the pad. The gap can be short or long. The longer the gap, the louder the water sound. Or you can make a cut in the pad that is just below the water.

When the water gets low, the gap between the pad and the water creates a sound when the water is running. The soothing sound acts as an indicator when the water in the holding area is getting low.

b. Changing the plants occasionally can make the greenwall appear to be a living, changing mural or artwork. Changing plants allows scents, shapes, and colors to vary along with the growth of the plants and can be easy to do if the wall is designed to allow for changes.

c. The space smells fresh and clean naturally. The fresh air can help with sleep, meditation, relaxation, ease and a sense of peace.

d. The plants' lights are daylight bulbs.

A FEW NOTES AS YOU ENJOY YOUR GREENING PROCESS

Greening your space can be a fun and creative endeavor. Any improvement, no matter the scale, helps. Resources for plants which can help remove contaminants can be found at www.engaging-green.com in the phytoremediation database link. Dream big and take it one step at a time. Over time you will see the totality of your design as wondrous and feel pride for all your accomplishments. The short-term and long-term changes, the improved air quality, the colors, scents and peace of living can be experienced and enjoyed daily. As you change, so does your design!

Chapter 2

Offices, Lobbies, Retail Space—Interior Commercial

INTRODUCTION

Green Solutions for Interior Commercial Greening

Creating a green space within a commercial building is one way that a company demonstrates that they value their employees and the greater environment. It attracts firms to rent one office space over another, impresses clients, and improves the quality of work life for all. Sharing space with multiple people in an office is an opportunity to portray the values and look of the business, while creating healthy spaces. Bringing in plant materials and green practices can improve the air quality, as well as increase employee health and well-being, retention, effectiveness, and long-term success for the individuals as well as the entire business.

Improving the quality of life in an office with green designs can be achieved through small design pieces as well as large scale works. The pieces can be created in stages using grey water from office sinks and by using full spectrum daylight bulbs. Adding healthy air and growing plants allows employees in a business to breathe easier and participate in environmental health with composting, grey water use, and other practices. Businesses, employees, and clients all benefit from a healthy, engaging environment.

Commercial sites have a great variety of design opportunities in both small and large spaces, transforming white office walls into colorful works of art, long hallways with many doors to large open spaces with much light-these areas offer opportunities to enhance all work spaces. The variety of commercial sites in this chapter includes lobbies, offices, and private retail sites. The application of the ideas shown can be useful for many sites with multiple users and changing needs, such as schools, hotels, manufacturing facilities, and factories.

Designs can include greenwalls, office dividers, planted artworks, mobile greenwalls and sculptures, and vertical and horizontal surfaces, and can all utilize grey water. The design ideas illustrated in this chapter can be applied to any site including schools, retail spaces, building lobbies, as well as service sites such as car dealerships, and spaces where employees and clients gather. Consider designs and materials that represent your business. Shapes, forms, scale, materials, and design details are as specific as the site itself.

Existing Site Conditions

This chapter delves into interior sites with multiple users and a variety of changing uses. With different cultures, ages, and multiple utilizations of the space, there are bound to be adaptations required of the green design. Creating mobile structures is an option which may be utilized. Additional design explorations can be employed for mixed age use, cultural references, hours of use, and to comply with different policies (for example, the Americans with Disabilities Act, ADA).

Designing for an office, business, or retail site means understanding how owners, management, employees, clients, and the public interact with the site. In most cases the office has many uses and styles, which changes as the business grows and sees a reduction or growth of employees, clients, and customers over a number of years.

Green practices can be applied to gathering spaces, conference rooms, eating spaces, multi-purpose rooms, mobile seating options, indoor parks, and mobile designs for improving cubicle and partition design, classrooms, meeting rooms, offices, work experience and education.

Commercial sites are likely to need more green design solutions than apartments and homes-not only due to the larger site scale, but also due to the quantity of people who use the space. Green designs can supplement existing engineered air quality systems. Air quality can be improved with green design and can also utilize grey water to further increase sustainability.

Green designs can improve the physical conditions of air, water, soil, and materials while simultaneously expanding the mental and spiritual health of the people who work in or visit the commercial site. Talking with a professional green designer about the specifics of your site is always a welcome first step.

WHAT TO DO NEXT

Questions to Ask about Your Site

In commercial sites design practices start with determining plants and other green practices which are best suited for the design. Plants can be utilized in greenwalls, hydroponic design, and aeroponic designs. Look for design treatments which can improve air quality, use grey water, add enough plants per square foot to clean the air of contaminants, remove the sources of contaminants, use local sources for materials, and add recycling and composting. Unique grey water sources such as water from air conditioners and dehumidifiers are also areas to explore in your site.

Take a walk through your site with fresh eyes or with someone who is unfamiliar with the site. Look at the lighting-is it healthy for plants or do you need to add daylight bulbs? What materials are around the office that may have unhealthy contaminants? These could include plastics, old sealants used in woods, carpets, machinery, fabrics containing chemicals, or cars that are driven into a showroom. What is in the space that is adversely affecting air quality and the overall health of the users and what can you do in the design to improve it?

Look at the materials of the desks, lights, chairs, walls, ceilings, walls, cabinets, blinds, cleaning supplies, presentation boards, and all other materials in the office. Are the windows leaking air, allowing in urban air or air with pollutants? Are the windows sealed well and keeping the indoor air indoors? If the air is sealed in well, then it is very important to create a healthy environment with no contaminants and good circulation.

Are the materials in good condition and maintained with natural cleaning agents, or are they disintegrating, cracking, or peeling? Peeling materials may be adding contaminants to the air. Consider whether replacing the cracked materials with newer healthier materials would be suitable in lieu of sealing the older materials. If one chooses to re-seal the older materials, look for sealants with low VOCs and for sealants that have a low chemical content to help improve air quality at the commercial site.

Consider what you can modify in both short- and long-term design and practices. Think creatively; use other chapters to help inspire you. To shed light on the application of green designs and working with site conditions, two sample sites are used with illustrated applications of green designs and practices. All sites have conditions which can be improved with green designs. These two examples include an urban office in New York City and a mixed retail/office site of a car dealership in Hicksville, Long Island. Ashley Kaisershot's (AK) questions continue throughout the chapter to further explain topics as they arise.

Example A—Urban Office in New York City, NY

An urban building

- The building was built in midtown Manhattan in 1915, with almost 30,000 square feet of rental floor space on each floor.
- In addition to the rental office space there are also elevators, corridors, restrooms, stairwells, etc.
- Located in a metropolis, the conditions of urban sounds and urban air quality exist.
- Smoking is not allowed in the building. Street smoking is a common occurrence in midtown.
- Urban air typically contains a higher concentration of contaminants and PM (particulate matter) such as SO_2 (sulfur dioxide), NO_2 (nitrogen dioxide), lead, carbon monoxide, and ground level ozone.
- Contaminants are a result of vehicle engine fumes, fossil fuels, industrial production, and natural processes such as windblown dust.

An urban office

- The office is located in a 12-story building, with approximately 7,000 square feet for the business and approximately 40 on site employees.
- Although the outer walls of the office have windows that can open, the unit contains individual offices including cubicles without direct sunlight.
- The two sides of this corner office have windows which face Broadway and 36th street two-major streets.

Images 2.1, 2.2.

Manhattan, New York, fifth-floor office in twelve-story commercial building in Herald Square, midtown Manhattan.

Image 2.1 by Kristi Blokhin/Shutterstock.com

Image 2.2 by Gagliard Photography/Shutterstock.com

- On the fifth floor, the light is pleasant and sunny during the daytime.
- The private offices on the business site are approximately 200 square feet.
- Clients visit the office frequently, typically meeting in the conference room and private offices.
- There is a maintenance company which comes to clean the office, water the plants, and empty the trash. There is an additional maintenance company for the wall planters.

Image 2.3.

Lobby of building during the holiday season.

Image by Heidi Solander

The materials in the office...

- Present are contaminants from off-gassing and decay of wood sealants in floorsmolding, and furniture, fabrics, curtains, furniture, clothing.
- Urban air comes through the ventilation and is circulated throughout the office. Urban air typically contains a higher concentration of contaminants and PM (particulate matter) such as SO_2 (sulfur dioxide), NO_2 (nitrogen dioxide), lead, carbon monoxide, and ground level ozone.

Knowing all these factors, the resulting design can improve <u>employee health</u>, <u>employee retention, lighting,</u> and <u>air quality</u>. Contaminants on this site originate from furniture, carpets, ceiling materials, sealants, paints, cleaners, and urban air. Additional design solutions can be used for <u>composting</u> and <u>recycling</u>.

Image 2.4, 2.5, 2.6, 2.7.

Fifth-floor office with images of the reception area, conference room, hallway, cubicle area, and private offices. Manhattan, New York.

Example B—A Typical Car Dealership with an Open Showroom, Service Area, and Open and Private Offices

A car dealership

- This property is typical of a new car dealership or one which has been recently renovated.
- The dealership contains a showroom, conference rooms, private offices, lounge areas, a service drive, and workshop.
- The total square footage of the building is approximately 18,000 square feet.
- The total square footage of the lot including parking is approximately 28,000 square feet.
- Car dealerships typically have large windows to showcase their vehicles. There is usually a large amount of natural sunlight along with a well-lit display of cars.
- This dealership is a tall one-story building.
- Plant materials are found on the exterior of the dealership within the boulevard between the sidewalk and the street.
- There are presently no plants on the inside of the building.

The interior space of the car dealership…

- Cars are driven in and out of the showroom weekly.
- Car emissions contain contaminants of hydrocarbons, carbon monoxide, carbon dioxide, NOx, formaldehyde, and NMOG (non-methane organic gas) which is a VOC (volatile organic compound).
- The service area, though open and ventilated in some facilities, is not as well ventilated in others.

Images 2.8, 2.9.

A typical car dealership in the United States. The dealership is recently built.

Car dealership with a parking lot, showroom, semi-private cubicles, private offices, lounge, service area, and workshop.

Images 2.10, 2.11.

Having an open-plan showroom means the offices, cubicles, and cars are all in the same area. This is inviting for customers. It also means the air quality is shared.

- In different seasons, doors and gates of the service area are kept closed due to weather conditions.
- Contaminants from off-gassing and decaying of wood sealants in floors, molding and furniture, fabrics, curtains, , clothing are present in this site.
- Many of the private offices do not have windows.
- The ceiling of the car dealership can be over 25 feet high and include ceiling fans and ventilation systems for air circulation.
- Heating and cooling are run through ventilation systems. The ventilation is affected by indoor car emissions.
- Doors to the dealership are continually opening for customers, meaning the air from outdoors is continually affecting the indoor air.

Images 2.12, 2.13, 2.14.

The private offices are located off the showroom floor. At this location, the service area is connected to the building. There are doors which separate the service drive from the full building.

The Green Design Can Improve

Knowing all these factors, the resulting design can improve air circulation with fans and can reduce the contaminants from the car exhausts, commercial carpets, outdoor air of the parking lot and service area in front of the building, cleaners, furniture, paints, sealants, service drive, and workshop. Additional design solutions can be used for the customer waiting area and employee lounge areas, while also utilizing composting and recycling methods to increase sustainability. Additionally, there is high wear and tear in commercial sites from the public use of the space. The wear and tear can be reduced by having a permanent interior planted green design built directly into the space. This can raise clients' experience of the space and improve the customer experience in the long run.

Images 2.15, 2.16, 2.17.

Manhattan office with greening designs

Shared central open space of the office

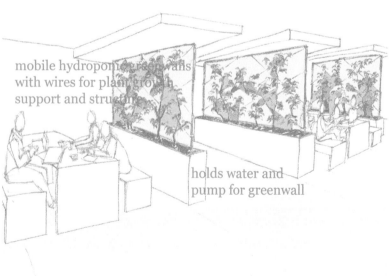

mobile hydroponic greenwalls with wires for plant growth support and structure

holds water and pump for greenwall

GREENWALLS AND DIVIDERS

This green office space design incorporates:

- Hydroponic greenwalls: walls which have plants growing with a water base rather than soil.
- Greenwalls with soil.
- A green alley with horizontal and vertical garden design and seating.
- A structure with irrigated plants.
- Grey and fresh water for use in the design.
- Compost and recycling functions.
- Daylight bulbs added throughout the office in the planted areas.

In the Manhattan office the plant choices range from plants for low light to high light. The use of artificial daylight bulbs increases plant health and also helps improve eyesight for those who come into contact with it.

Some of the design elements have their own water, pump, and lights set on a timer. This office contains its own grey water which can be used for watering the planters during the week as well as the weekend.

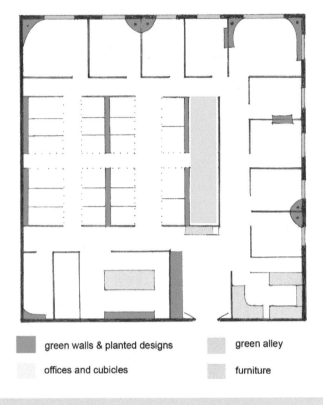

■ green walls & planted designs		■ green alley
□ offices and cubicles		■ furniture

Image 2.18.

This floor plan for the midtown office includes areas for planted green designs throughout the space.

Offices and commercial sites, no matter the age, the quantity of employees, or the scale, all contain sources of air quality contaminants. This is due to natural processes and off-gassing from objects and materials on the site. Contaminants may be found in such basic places as:

- Sealants used for wood
- Paints
- Cleaners
- Office materials and copy machines
- Unclean air ventilation systems
- Floor sealants
- Heating systems
- Office kitchens
- Outdoor air containing car exhaust fumes and other contaminants coming in offices through windows

Interior green designs improve quality of life in an office. Through multiple greening designs and practices, you can enjoy many benefits.

Designs to increase sound absorption in open offices can be another way to positively impact an office environment. Acoustics in open offices can be improved with green designs which help define spaces and sub-spaces, with both mobile and fixed designs. An indoor space designed with plants and seating can act as a healthy gathering space. Light quality can also be improved through the use of daylight bulbs which are used for plants. A full daylight spectrum improves eyesight and tired eyes from overuse of computer screens.

Green designs can also improve air quality with the use of specific plants to reduce contaminants in the air through phytoremediation-the use of plants to uptake contaminants from the air, soil or water. Sick building syndrome is reduced with green designs. Further information about SBS can be found in local and regional research.

The ability to increase mental well-being through green office environments are another benefit of implementing green designs. People experience stress reduction through visual and physical interaction with green designs. Gardening, interior or exterior, can help reduce stress. People can see time change through the color change, growth, and flowering of plants. Sound features with water elements can create an enjoyable atmospheric sound.

Employee and student productivity are also increased with interior green designs. You can improve sinus health with plant scents and fresh air, as well as improving one's mental health and physical health, therefore reducing sick days. For students specifically, studies have shown that viewing and exposure to greenery while studying or researching improves retention of knowledge.

Additional benefits include aiding natural bacteria and germs. In an office setting, using recycling and compost areas allows employees to contribute to a greater environmental cause. The use of grey water from office sinks and dishwashers can be recycled for green design, reducing fresh water needs.

AK: *What is **ADA**, and what does it mean in design?*

SF: ADA stands for Americans with Disabilities Act. This applies to the design of offices and public settings, to create accommodations for a variety of disabilities. The most common disabilities include visual impairments, walking impairments, and height impairments. These designs are not only helpful to those with physical disabilities they also improve the site for users of diverse ages and mobility.

Designing for ADA standards is not just a legal requirement, it is a good practice for all designers to be inclusive of diverse and changing users with integrated green designs.

*AK: How do you **design for the needs of diverse people** including those with disabilities?*

SF: Designing for people with disabilities includes designing to create an experience that is similar for all users. Integrated designs for people with walking or visual impairment allows all users of a space to work together, rather than creating a separate or inferiordesigned space for any person.

There are some basic matters to consider in successful design, such as ease of movement. Making sure all the walking areas are clear and wide enough for users is a factor to integrate into the site design. Additionally, integrating signage and good lighting makes areas unambiguous for users. It may be useful to work with design professionals to explore options for your site.

Image 2.19.

As in all greenwalls, plants can be replaced as needed. The plants chosen throughout the green design in the Manhattan office are for air quality improvement as well as for benefits of the employees.

The plants include:

<u>For formaldehyde and carbon monoxide from the furniture:</u>

Spider Plant (*Chlorophytum comosum)*
Chrysanthemum (*Chrysanthemum x morifolium)*
Orchid (*Dendrobium taurinum)*
Warneckei (*Dracaena deremensis 'Warneckei')*
Red-Edged Dracaena (*Dracaena marginata)*
English Ivy (*Hedera helix)*
Lady Palm (*Phapis excelsa)*
Snake Plant (*Sansevieria trifasciata)*
Peace Lily (*Spathiphyllum wallisii)*

The nutrients needed for plant growth are put in the water system. These nutrients can be bought by the gallon and are used on a schedule based on the amount of plants and water. Typically, a small amount of nutrients are put in the water once a week or bi-monthly. An exact determination of the quantity is based on the scale of the wall and the amount of water used. Water naturally evaporates in a space and must be replaced in the planted designs.

Image 2.20.

The plants for air quality include:

<u>Plants for remediating Polycyclic Aromatic Hydrocarbon (PAH):</u>

Buffalo Grass-
   (*Bouteloua dactyloides*)
Canadian Wild Rye-
   (*Elymus canadensis*)
Red Fescue-
   (*Festuca rubra*)
Ryegrass-
   (*Lolium multiflorum*)
Empire Bird's-foot Trefoil-
   (*Lotus corniculatus*)
Alfalfa-
   (*Medicago sativa*)
Western Wheatgrass-
   (*Pascopyrum smithii*)
Reed Canary Grass-
   (*Phalaris arundinacea*)
White Clover-
   (*Trifolium repens*)

AK: *Does switching from **chemical cleaners to organic solutions** help with the air quality? And do they work well?*

SF: The cleaners you use do make a difference in air quality and using cleaners that are effective for the task is something we all look for. Whether you research making your own organic cleaners from natural ingredients (for small offices that do their own cleaning) or you ask your cleaning staff to switch to organic solutions, the effects occur rapidly.

Common cleaners contain approximately 62 chemicals.

Some of the chemicals found in cleaners include phtalates, perchloroethylene (perc), triclosan, quarternary ammonium compounds (quats), butoxyethanol, ammonia, chlorine, and sodium hydroxide. These chemicals have been linked to headaches, skin burns, asthma, respiratory disorders, reproductive disorders, hormone disruption, neurotoxicity, and cancer.

There are recipes for cleaners which can be used in lieu of chemical based ones. Some natural ingredients include baking soda, vinegar, tea tree oil, peppermint oil, lavender oil, and other essential oils that can be found in local natural food shops.

AK: What are **daylight bulbs,** and do they really make a difference?

SF: Bulbs with lower degrees Kelvin than daylight have yellow, orange, or green tints. Look at a building with lights on at night to see the variety of bulb colors there are that are different than daylight.

Daylight bulbs have a full light spectrum to make a pure white light and are easier on the eyes. Using daylight bulbs throughout a space, in addition to the greenwalls, can alleviate eye strain and discomfort.

Full spectrum light uses all the colors of sunlight-which is measured at 5,600–6,000 degrees Kelvin. Lights which appear yellow, green, or other shades other than pure white are less than 5,600 degrees Kelvin. Full spectrum light is healthy for humans as it allows them full range of eye muscles. It is also healthy for plant growth. Ottlite and many other companies produce daylight bulbs.

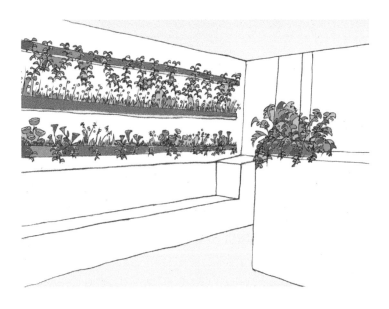

OFFICE ENTRANCE

Image 2.21.

This is an opportunity for an office to make a statement and more positively affect how each employee starts their day and finishes it. The entrance can be bold, beautiful, and be part of the gathering space for employees.

In this New York City office, a series of greenwall pieces are placed throughout the entrance and continue throughout the entirety of the office. The entrance has the boldest colors and fresh scents in a piece which is inspired by the planning and zoning patterns this company does.

PLANTS FOR THE ENTRANCE DESIGN

With the addition of daylight bulbs added to the entrance, some plants which would be effective for this design include:

For colorful blooming aesthetics:

Flowering Maple (*Abutilon*); Lipstick Plant (*Aeschynanthus radicans*); Laceleaf (*Anthurium*); Begonia- (*Begonia*); Calamondin Orange (*Citrofortunella microcarpa*); Kaffir Lily (*Clivia miniate*); Crown of Thorns (*Euphorbia milii*); Geranium (*Geranium*); Guzmania (*Guzmania*); Hibiscus (*Hibiscus*); Ixora (*Ixora*); Jasmine (*Jasminum*); Guppy Plant (*Nematanthus nervosus*)

HALLWAYS

Image 2.22.

The hallways of the office can be mixed with traditional images as well as living designs. The living pieces can be as different as the framed pieces. Examples of groupings of plants could include annual herbs, grasses, or succulents. Having a variety of sizes, shapes, styles, plants and lighting can change the monotony of walking down a hallway into walking down a changing gallery of art.

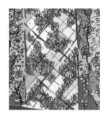

AK: *Can I **phase the design** in stages because of budgets?*

SF: Absolutely! Phasing in designs in a site is common and expected. Creating a master plan is very important. A master plan reflects the full scope of the design and states the needed electricity, water, lights, and other aspects of the complete project without sacrificing a healthy environment. Creating the phases which lead to the final plan over a series of months or years (depending on the scale of the site and design) helps the project to be completed on time.

Choose 2–4 stages to the green design process which work with the employer, employees, and the budget.

HALLWAYS

Designs or plants for the hallways with low natural light and added daylight bulbs can include:

Chinese evergreen (*Aglaonema*); Peace Lily (*Spathiphyllum*); Spider Plant (*Chlorophytum comosum*); Lucky Bamboo (*Dracaena braunii*); Bar-Room Plant (*Aspidistra elatior*); Devil's Ivy (*Epipremnum aureum*); Philodendron (*Philodendron*); Painted-Leaf Begonia (*Begonia rex*); Laceleaf (*Anthurium*); and a variety of ferns

Images 2.23, 2.24.

A car dealershipgreen design throughout parking lot, showroom, service areas, lounges, and offices.

GREENWALLS AND PLANTERS

Throughout the Car Dealership

Car dealerships are large in scale to allow for cars to be displayed, stored, moved, and cared for on site. With such a large space, creating an overall master plan for a green design can help unify a space. In addition to the aesthetic appeal, the plants can help support the engineered ventilation system as well as improve air quality and likely use less energy.

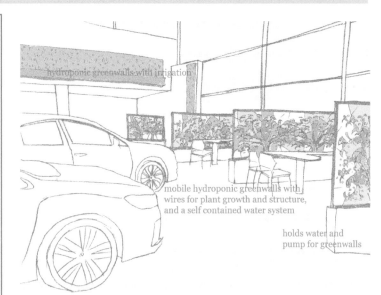

The plants for this design include plants which love light and address the contaminants from car emissions.

<u>Plants which uptake car emission contaminants include:</u> Spider Plant- (*Chlorophytum comosum*) Chrysanthemum- (*Chrysanthemum x morifolium*); Warneckei (*Dracaena- deremensis 'Warneckei'*); Red-Edged Dracaena- (*Dracaena marginata*); English Ivy- (*Hedera helix*); Lady Palm- (*Phapis excelsa*); Dwarf Date Palm- (*Phoenix roebelenii*); Snake Plant- (*Sansevieria trifasciata*); Peace Lily- (*Spathiphyllum wallisii*);

Image 2.25.

Car companies market more than the cars: they market a lifestyle. The lifestyle may include ease, freedom, comfort, safety, health, and more. Accordingly, the dealerships are representative of that lifestyle. Having a green dealership which is inviting and has a healthy environment improves the lives of many people, as well as promoting a healthy company and lifestyle.

In creating a healthy space, car dealerships have many of the same elements as an office, with the additional design opportunities of a much larger space: multiple floors, private offices, customer areas, daily customers, a working garage, parking lots, doors being opened routinely, storm water, and car exhaust fumes. Creating a variety of green planted sculptures, wall planters, recycling areas, composting areas, and ways to utilize grey water and storm water can be an asset to the company. The green practice helps with customer relations, media and press opportunities, employee retention, and offers tax credits in some states.

On this site, multiple designs and practices are created throughout the building through the use of mobile and static designs. In the showroom there are mobile green pieces on wheels, each with its individual water container and light. They are mobile to accommodate the routine changing of car displays. The customer lounge has permanent planted pieces, along with grey water and storm water systems for the plants. Recycling and composting areas are also built into the customer lounge. Sales offices and managerial offices have smaller pieces which relate in aesthetic to the larger pieces, but are scaled appropriately to the space. The drive-in service area has effective ventilation and plants which reduce the contaminants from car exhausts. Though each piece is individually monitored for water, light, and electricity, the aesthetic form of the pieces have elements which visually create a cohesive piece.

AK: *Is there a reason to create a **mobile design versus a static design***?

SF: This is purely based on the needs of site. In some cases a static design makes more sense, such as at the entrance to the office or business, in conference rooms and private offices. Mobile designs such as greenwalls or planters on wheels are best used when the common area or lounge area needs to be frequently adapted for events, office growth, or the needs of the business. In the car dealership, movable pieces for the showroom make sense as the cars are moved routinely. Permanent pieces on the walls also add to the overall design to create anchored pieces for offices, entrances and lounge areas.

AK: Do **different types of office and retail companies** *have different contaminants and design solutions?*

SF: Yes and no. There are many variables and each site must be looked at individually. However, in these two examples, the businesses have some similar contaminants (furniture, wood), and vastly different contaminants (car exhaust fumes). There are similar sources of contaminants in businesses such as furniture, sealants, paint, and cleaners. There are also differences in each business, arising from the architecture, location, business type (car dealership, clothing retail, building supply unit, dry cleaners, bakery, etc.) which create different environments and need to be looked at individually.

CUSTOMER AREA WITH FOOD AND DRINKS

Image 2.26.

Analyzing this space allows one to see where green designs can be placed, as well as sources of grey water. In this space, grey water sources are marked in blue. Grey water can be collected in tanks below the sinks, as well as below the coffee machine where the coffee spills. Grey water calculators are commonly available on websites. Greenwalls can be large, small, and uniquely shaped. In this space, the opportunities to utilize greenwalls are indicated in light green shapes.

INDIVIDUAL OFFICE IN CAR DEALERSHIP

Image 2.27.

A planted greenwall is behind the desk. The client experiences the view, scent, and air improvement benefits of the planted piece in the short term, while the office employee experiences long-term improvements.

Image 2.28.

Greenwalls add a fresh scent, as well as soothing water sounds in the customer lounge area.

Image 2.29.

<u>In the service drive-in area, additional plants which address car emissions include:</u>

Buffalo Grass (*Bouteloua dactyloides*)
Blue Gamma Grass (*Bouteloua gracilis*)
Bamboo Palm- (*Chamaedorea seifrizii*)
Orchid (*Dendrobium taurinum*)
Red-Edged Dracaena (*Dracaena marginata*)
Rubber Plant (*Ficus elastica*)
Weeping Fig (*Ficus benjamina*)
Gerbera Daisy (*Gerbera jamesonii*)
Boston Fern (*Nephrolepis exaltata*)
Kimberly Queen (*Nephrolepis obliterata*)
Heart-Leaf Philodendron (*Philodendron scandens*)
Dwarf Date Palm (*Phoenix roebelenii*)

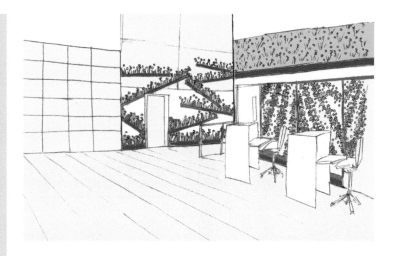

AK: *Do I need to consider **bugs or insects** with interior planted designs?*

SF: Though insects or bugs can be attracted to plants, there are some easy and natural solutions to help your design stay free of insects. The first thing to note is, if you are not in an area where the plants are exposed to outdoor access bugs will not get into the space. If you are near a doorway which opens frequently and bugs do come into the space, there are natural solutions such as using little or no soil. Without soil, insects do not have an area to live. A water reservoir is designed to keep water aerated, that reduces long-standing still water.

There are also natural solutions to use if insects or bugs arise. Natural solutions include:

Soapy water that dehydrates aphids and spider mites.

Pepper sprays that contain capsaicin (black pepper, chili pepper, dill, ginger, and paprika).

Citric acid (orange juice, lemon juice), sage, thyme, basil, rosemary, mint, rue and lavender can be added to water to repel or kill some insects.

Pyrethrum spray is made from chrysanthemum flowers and paralyzes flying insects on contact. A clove of garlic placed in the soil of plants keeps bugs away.

Images 2.30, 2.31.

A lobby design: green designs act as artworks as well as vertical and horizontal gardens.

This large open lobby has large windows providing heat and hours of natural sun.

Whether designing for a large open lobby in a residential or commercial building or a small lobby in a mixed-use building, these designs set the stage for the building. Think of the times you have been greeted by a door person with a smile as you enter a building. Having fresh air, fresh scents, and changing greenery can help greet all entrants to the building, as well as act as a gathering space for people. Additionally, it can add financial value to the building to have a green lobby.

Each piece has its own timer, lights, pump, and water system. Grey water from the building is used in this design.

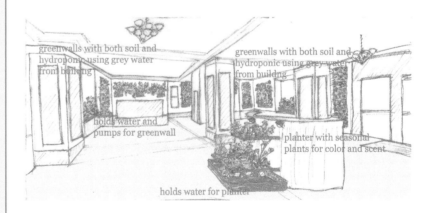

A combination of vertical and horizontal designs, along with hydroponic, soil, and aeroponic pieces are created for a variety of textures, colors, and blooms throughout the year.

Plants for colorful blooming include:

Oxalis (*Oxalis*); Brazilian Fireworks Maracas (*Porphyrocoma pholiana*) African Violets (*Saintpaulia*) Streptocarpus (*Streptocarpus*)

TWO-STORY LOBBY GREENWALL OR SHOWROOM

Image 2.32.

The design is based on an existing New York City showroom with a large window front. The concept is applicable to any retail showroom, lobby, or retail area with large windows.

The large greenwall is divided into three sections. The middle section has patterned wires for vines or air plants to grow. The wires can be in a pattern specific to the company or architecture. The outer panels have plants in soil, allowing for a more diverse planting.

The irrigation system is on a timer, with water piped directly from the reservoir when needed. A pleasant gentle water sound can easily be incorporated into this greenwall in the wired section or the outer panels.

AK: What is **engineered soil or structured soil**?

SF: Engineered soil or structured soil is soil that has been amended to account for specific challenging conditions such as weight or compaction. For example, soil under sidewalks is typically compacted so the walkway does not shift. However, compacting the soil does not allow for street trees and other plants to grow roots naturally due to the compaction. Another instance when structured soil is beneficial is where the weight of the soil is important, such as on roof designs. Structured soil is lighter than traditional soil and can be used in cases where the weight is a factor. There are companies which produce structured soil based on the site condition. Information on engineered soil can be found at your local and regional nurseries.

MULTI-STORY LIVING WALL AT THE UNIVERSITY OF GUELPH HUMBER, GUELPH, ONTARIO, CANADA

Image 2.33.

Quoted text excerpts on this project are from Architectural Record article *Indoor Air Biofilters Deliver Clean Air Naturally. Biological Systems function to improve air quality while providing beautiful form* by Peter J. Arsenault, FAIA, NCARB, LEED AP, and Alan Darlington, PhD.

The living wall at the University of Guelph was installed in 2004. The multi-story greenwall is approximately 33 feet wide by 52.5 feet high, with 525 square feet of living plants. This wall is supplied by natural light and supplemented by artificial light.

"Indoor air biofilters provide a full range of contributions toward greening a building. The indoor air biofilter can be thought of as a system of exposed plants that are integrated within a building and appear as a vertical hydroponic wall."

Image by Steven Evans Photography

THE SCHEMATIC MAKE-UP OF AN INDOOR AIR BIOFILTER

Image 2.34.

Living Walls: Nedlaw Living Walls

"Behind the scenes, a pump constantly circulates water from a reservoir at the base to the top of the wall. The water then flows down the wall through a porous synthetic root medium in which the plants are rooted. Air from the occupied space is actively drawn through the plant wall by either the HVAC system or onboard fans and then returned to the occupied space. As the dirty air from the space comes in contact with the growing (rooting) media, contaminants move into the water phase where they are broken down by the beneficial microbes growing on the roots and other surfaces in the growing media."

Image by Diamond Schmitt Architects

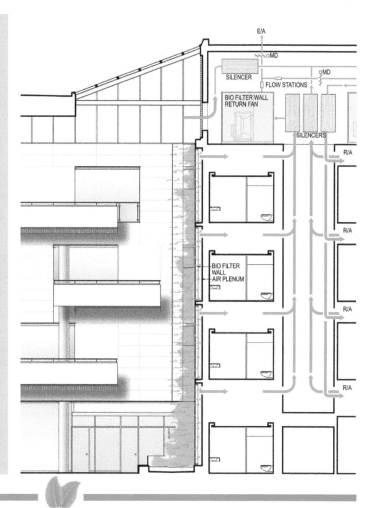

AK: *Why is choosing **local sourcing** something I should consider?*

SF: Choosing local sourcing works on a few different levels. One level is that local sources do not have to commute or ship their product from long distances. This creates a lower carbon footprint for their product or service and may cost less. Additionally, local sources support local businesses and keeps community businesses intact and growing. It is likely that your business was once local or small. Local sourcing allows the business to develop a reputation while growing towards a larger field. Though prices of local or smaller businesses may be a bit higher, using a business with a local representative who can come to your site to offer a personal service while simultaneously lowering the environmental impact of moving many supplies long distances is an approach to take for long-term success. Look for companies who have a short commute and use organic materials or low impact as applicable.

LIVING WALL AT UNIVERSITY OF GUELPH

Image 2.35.

"An indoor air filter to breaks down VOCs through the biofiltration process. In controlled laboratory studies, these systems have been shown to remove up to 90 percent of VOCs including formaldehyde, in a single pass. Real world testing in actual buildings shows some expected variation from ideal laboratory conditions, but they have still been measured at impressively high VOC removal rates."

Image by Elizabeth Gyde, Diamond Schmitt Architects

AK: *What **quantity of plants** is needed to be effective?*

SF: This is a bit complicated for a short simple answer. One researcher, B.C Wolverton, has studied plants to determine the amount of plants which are effective in an interior space. He has been doing studies for NASA for decades. However, there are multiple factors to consider in addition to square feet and cubic feet.

The size of a plant, effectiveness of particular plants, clumping of the plants, air circulation, and other factors affect the calculation of plants needed for a space. Another factor to remember is that plants grow and will need larger pots over time. This is healthy, though also a factor in determining how many plants are needed in a space.

With all that, the images shown in this book use a quantity of plants which will be most beneficial for the spaces shown. Using a variety of plants to help with different air contaminants is also beneficial. The addition of green design, no matter the scale, is always beneficial.

"Indoor air biofilters also contribute to removal of particulate matter in the air. By contrast, an indoor air biofilter uses the extensive surface of the plant material to significantly reduce airborne dust particles in addition to reducing chemical contaminants. This means that the system works very effectively as a total air 'pre-filter' before the processed air enters or returns to the HVAC system. Dust particles collected in the biofilter are then washed away by the flowing water cascading down as part of the hydroponic process."

Image 2.36.

"Buildings with biophilic-inspired elements have been demonstrated to provide psychological benefits to their occupants in a variety of ways. Studies show that workers in offices with views to nature tend to feel less frustrated, more patient, report higher levels of overall satisfaction and well-being, and are therefore more productive employees. These benefits coupled with the indoor air quality improvements lead to decreases in common office ailments (e.g., less fatigue and fewer headaches, sore throats, coughs, and dry skin issues), decreases in employee absenteeism, and typically increased employee productivity. One of the more difficult human resource challenges is actually not 'absenteeism' but termed 'presenteeism,' which is lost productivity as people mentally 'check out.' An improved workplace with a varied environment and sensory stimulus allows for people to be happier, more relaxed, and stay 'tuned in' longer."

Image by Elizabeth Gyde, Diamond Schmitt Architects

A FEW NOTES AS YOU ENJOY YOUR GREENING PROCESS

Green designs and practices have many benefits which help businesses achieve success for the people that work there as well as clients who visit the office. Long-term business value can be seen in the design of the company spaces.

Create designs which have your unique form and that work with the function and needs of the space. Form and function working together are a great asset in any design. As you develop your ideas, there are opportunities to work with green professionals to help you bring your ideas to life. Some professionals you may consult include interior designers, contractors, nurseries, landscape architects, architects, lighting designers, horticulturalists, soil scientists, and green design specialists.

This chapter is intended to inspire you to realize that greening can be unique, effective, engaging, and profitable. Working with stakeholders and professionals as a team in a collaborative effort is the goal of all projects. Green design reflects a corporate value and can proudly be an element of the branding of a company as positive and progressive.

Chapter 3

Houses, Residential Sites—Interiors and Exteriors

INTRODUCTION

Green Solutions for Residential Interiors and Exterior Greening

A home is as unique as the people that live in it. In this chapter the homes range from connected urban houses, sprawling suburban houses, to unique buildings made into residences. The ideas in this chapter can be applied to many styles of houses in different areas. Interior and exterior spaces can be used as opportunities for green designs on a variety of scales.

The user of the site is of great importance in house design. The users can be families with a great range of people, pets, and guests. Creating a design which is beautiful while improving the quality of life is the goal.

Greening can create fresh air, collect and reuse grey water and storm water, have applications with scented oils, color accents, recycling, composting, erosion control, create indoor/outdoor rooms and spaces, dampen unwanted sounds, and frame and block views. The designs can be vertical, horizontal, and any angle in between, with both small-and large-scale pieces. It is the combination of all these factors that makes each site have opportunities that are unique.

Designs can include kitchen gardens, vertical and horizontal designs, hanging planted sculptures which slide across windows acting as art while also hindering views, a piping system for the grey and storm water adding a visually interesting element to room design, permeable pavers in driveways for slowing and cleaning storm water, interior and exterior composting units, hydroponics, aeroponics, unique garden solutions, and enough plants to clean the air contaminants and contaminants from local sources.

Existing Site Conditions

With this chapter being one that explores both interior and exterior sites, green design allows for a unique blurring of lines as to what can be experienced as interior/exterior, and shifts the perspective of both. Additionally, there are lighting, scale, grey water, and storm water opportunities to consider for houses of all types. Designs allow for the reuse of water, reducing the need for fresh water for plants, while improving air and soil quality through design. Looking at the existing site conditions one can identify grey water sources and get a better understanding of grey water and its uses, as well as using slopes and design to improve water management. Additionally, looking at rain water harvesting may be an opportunity on your site.

Understanding the existing conditions of your site leads to successful design solutions. Let's break down a few elements to look for in your site:

Site conditions which may be challenging,

- Understanding the sunlight and other sources of light are integral for a successful design. As with all structures, looking at areas which get more and less natural light will allow you to consider if additional light sources are needed for plants.
- Slopes can range from gentle to severe across a lot. Look at the slope of your walkways, roof, yard, seating areas, driveways, etc. Are there areas which puddle up frequently with rain? Are there areas which are easily eroded due to a severe slope and unprotected soil?

- Contaminant sources—where and what are they? Look at the materials you have around your house—the walls, furniture, floors, ceilings, roofs, walkways, items in garage storage, cleaners, or sprays. Are there some items you may want to remove on account of what they are made of?

Site conditions which can be used in design:

- Grey water—look for all sources of grey water. Grey water typically comes from sinks, showers, dishwashers, and washing machines. It is not the water from toilets, that is called black water. How many gallons of grey water do your produce a day? There are many resources for grey water calculations online.
- Rainwater and storm water—what is the average yearly rainfall in your area? Can you capture it and use it on your site for an outdoor shower or watering the plants? There are multiple resources for rain and storm water charts online.
- Composting and recycling—what do you have around the house that can be recycled or put in a composter? Look for items which you can recycle and compost. As you begin to notice your choices, you may choose to change some of them. For now, work with what you and your family typically process in a week and create a design which allows for those items.

Take notes about all of these on your site and work with what you have. What you do with this information will inform your design choices and with this knowledge you can have fun while improving the quality of life for you and your family.

WHAT TO DO NEXT

Questions to Ask about Your Site

Once you have looked at your home and evaluated a typical week in it, the next step is to have fun dreaming up ideas. There are many places throughout the site to consider in your design. Start by making a list of all the areas throughout the site that may be used for design, and a second list of all the specific elements of your site that you want to address. Do not filter your list based on budget or time. These lists are your dream lists and may be fantastically wild and long. Your lists may look like this:

Areas for green designs
- Notably pitched or flat roof top
- Walls, floors, ceilings, vertical surfaces, horizontal surfaces
- Attic with sunroof
- Artist's studio
- Main bedroom
- Guest room and office
- Bathroom
- Hallway
- Stairway
- Kitchen
- Living Room

- Dining Room
- Detached garage, including rooftop
- Front porch
- Back deck
- Fences—front yard and back yard
- Driveway
- Walkways
- Area with garbage and recycling cans
- Front yard
- Back yard
- Pond

Areas to address through green design
- Composting
- Recycling
- Car exhaust fumes, especially in the garage
- Gas stove contaminants
- Contaminants from off-gassing and decaying of wood sealants in floors, molding, and furniture
- Contaminants from off-gassing and decaying of fabrics, curtains, furniture, clothing
- Indoor kitchen herb garden
- Contaminants from cleaners
- Contaminants from beauty products (bathroom sprays)
- Grey water reuse
- Storm water reuse
- Rainwater garden
- Improved heat and air circulation
- Block view of a dog play area in the yard next door
- Block view of neighbor's television shining through the curtain late at night
- Gentle sound from a gentle water feature
- Scent of fresh greenery and fresh air inside, all year round
- Counter the effects of the winter season with indoor full spectrum lights and blooming
- plants

After you have made your own list, start putting the list in order of priority, including which ones you want and can address first. Make some short-term and long-term plans. In a house it is best to consider temporary and mobile designs along with long-term built-in designs. Mobile designs can be considered green artwork. Some paintings or sculptures are easy to move and rehang, while others are anchors for other designs to move around. The same applies to interior and exterior planted areas. Creating some designs which are small, hanging, on wheels, or lightweight can be useful for changing designs as the needs of the site change.

Composting is beneficial for improving soil conditions. There are many items that can be composted to add nutrients to the soil. This list is edited from the website www.smallfootprintfamily.com and shows some items which can be composted.

From the Kitchen

Fruit and vegetable scraps	Egg shells, crushed	Coffee grounds	Coffee filters
Tea leaves, loose	Tea bags, of natural materials	Spoiled soy/ rice/ almond/ coconut milk	Stale beer and wine
Paper bags, shredded	Crumbs you sweep off the counter	Cooked pasta	Cooked rice
Stale breads, pittas, and tortillas	Spoiled tomato sauce	Stale crackers	Stale cereal
Stale pretzels	Spoiled tofu	Seaweed, kelp, nori	Avocado pits
Old herbs and spices	Old jelly, jam, preserves	Old oatmeal	Unpopped and burnt popcorn kernels
Stale pumpkin, sunflower, and sesame seeds, chopped up	Nutshells (except walnuts, which are toxic to plants)	Pizza boxes, unwaxed, cut into small pieces	Cardboard egg cartons, cut up
Wooden toothpicks	Bamboo skewers, broken into pieces	Natural wine corks	Dry dog or cat food, fish pellets

From the Bathroom

Hair from hairbrushes and electric razors	Toilet paper roll, shredded	Loofahs, natural, cut up	Used facial tissues
Cotton from cotton balls and cotton swabs			

From the Laundry Room

Dryer lint, from 100% natural fabrics	Old cotton clothing and jeans, ripped into small pieces	Cotton towels and fabric scraps	Old wool clothing, ripped into small pieces

From the Office

Bills and other plain paper documents, shredded	Envelopes, shredded, without the paper windows	Pencil shavings	Sticky notes and business cards (not glossy), shredded

From around the House

Dust bunnies from wood and tile floors	Newspapers, shredded	Junk mail, shredded, without plastic coating	Old rope and twine, natural, cut up
Dead leaves from houseplants	Dead houseplants and soil	Flowers and floral arrangements	Dead seasonal leaves
Used matches	Grass clippings	Natural potpourri	Burlap sacks, cut up
Sawdust, from plain wood that has not been treated, stained, or painted	Ashes from the fireplace, barbecue grill, or outdoor fire pits (in moderation)		

From Party and Holiday Supplies

Wrapping paper rolls, cut up	Paper table cloths, cut and shredded	Crepe paper streamers, cut up	Jack o' lanterns, smashed
Hay bales used at fall décor, broken apart	Natural holiday wreaths, cut into pieces	Christmas trees, cut up	Evergreen garlands, cut up

It is best to consider the effects of the whole design when looking at your home. Look at all the possibilities of a healthy environment, both short-term projects and long-term design. Dreaming up the ways in which you wish to change your lifestyle to an easy green lifestyle can and should be fun. The choices include but are not limited to composting, recycling, pure spectrum lighting, reuse of grey water, supporting local businesses, choosing organic, adding diffusers for scented oils, hanging aeroponic artworks, adding greenwalls, tables with herbs, living artworks, and more. There are no limits.

Designing green is about adapting your home for inspiration, safety, health, love, and growth. Your designs will change and grow as your style and sensibilities change. Living in a healthy space may inspire subtle changes to your designs, or drastic ones. Consult an expert when you need, and always have fun.

The next two examples apply some of these ideas to an existing site. The two examples include an older, larger two-story home with a yard in the Midwest, and a house in an industrial town which is at the end of a row of terraced houses. Ashley Kaisershot's (AK) questions in this chapter develop further understanding of the written and illustrated topics.

Example A—Two-Story House with an Attic, Basement, and Separate Two-Car Garage in the Suburbs, Fargo, ND

Exterior and general information...

- The house was built in 1903 on a lot which is approximately 5,894 square feet.
- The two-story house is approximately 1,616 square feet. There is also a basement and attic which make up the full footprint of the house, as well as a detached garage.
- There have only been five owners of the house since it was built.
- Currently the owners have no children, two cats, and frequent guests.
- The major material of the exterior of the house is painted wood.
- In 2012 the house was repainted, and the exterior lead paint removed.
- Winter in Fargo, ND is infamous. Notoriously, the temperature routinely drops below 0-degrees Fahrenheit.
- Snow drifts on the property create a wall of snow in the driveway and along the fence.
- Fargo is notably flat. The estimated less than 1% slope from the front sidewalk of the property to the edge of the backyard is subtle and inconsequential.
- The fence at the side of the house is wooden on the side and at the back there is a chain-link fence. The vines are thick from June–September. The other times of the year the backyard can be seen by the back neighbor.

Image 3.1. Two-story house in Fargo, ND. The house has had renovations added by different owners, including a new porch, a back deck, an added garage and the removal of exterior lead paint.

Image by Tracy Green/Green Team Realty

Images 3.2, 3.3, 3.4. The tandem garage has a walkway to the back of the house and the backyard. The window at the back of the house was removed during an expansion of the kitchen.

The front room in the house is a living room/parlor room with a large window.

Images by Tracy Green/ Green Team Realty

Interior rooms...

- The windows of the house have their original frames, and most have the original single-pane warped glass which are beautiful but admit air, affecting the house temperature. Sealing the windows in the winter is part of the seasonal upkeep.
- With large windows on three sides of the house, there is sunlight during most of the day throughout the year.
- The neighbors on one side play their television loudly and have held parties which create sound and light at the side of the house at night.
- Radiators heat each room, and air conditioners are put on each floor to cool the house in the summer.

Images 3.5, 3.6. The hallway leads to the first floor and the main staircase to the second floor.

Images by Tracy Green/ Green Team Realty

Images 3.7, 3.8, 3.9, 3,10.

On the first floor are the living room/parlor, dining room, kitchen, and half bathroom/laundry room. There are stairs to the basement from the kitchen. The basement is the full footprint of the house.

Images by Tracy Green/ Green Team Realty

- There is still interior wall paint that has lead in it, though it has been painted over.
- The original wood floors, molding, and doors of the house were once painted, then the paint was removed and the wood sealed with a polyurethane sealant. Polyurethane sealants contain isocyanates, amines, glycols, and phosphates. Isocyanates have been shown to cause skin and respiratory sensitivities, while the other chemicals risk causing similar issues.
- The attic is used for storage and has rolled fiberglass insulation across the floor to help maintain the temperature of the house. Exposure to fiberglass insulation has health risks when in contact with skin or inhaled.
- The basement is full ceiling height and the full footprint of the house and currently acts as storage. It was once designed as a cool room for the summer season and safe space during storm seasons, though it is used for storage in the present day.

Images 3.11, 3.12, 3.13, 3.14.

On the second floor there are four rooms—a master bedroom, guest room, office, and family bathroom. There is a staircase from the office to the attic.

Images by Tracy Green/Green Team Realty

Knowing all these factors, the resulting design can improve the psychological effects of the long winter season with <u>interior vertical blooming designs and exterior painted snow</u>. The air quality can be measurably improved through designs with plants creating fresh air to counteract the contaminants from the <u>wood</u>, <u>furniture</u>, <u>curtains, household cleaners</u>, <u>paints</u>, <u>cabinets,</u> and <u>exterior air</u> coming in through the windows and open door. <u>Air circulation</u> to improve heating and cooling throughout the two main floors of the house can be created with a combination of ceiling and wall fans to move air. Designs can <u>diffuse the sound and light from the</u> neighbors' television. Additional design solutions can be used in the <u>garage for the car exhaust fumes,</u> exterior <u>driveway</u>, <u>rainwater</u>, <u>grey water,</u> <u>composting</u> and <u>recycling</u>.

Example B—Two-Story Connected House Which Is at the End of the Structure in the Industrial Town of Wolverton, United Kingdom

An attached house…

- The residence was constructed between 1900–1910 and is the end house in a series of connected Edwardian houses. Across the street is a tall brick wall blocking a railway station.
- It is a two-story building of approximately 1,233 square feet, with a sloped roof, high ceilings, brick walls, and wood floors.
- There is a front entrance, front garden, back deck, rear garden with an outdoor water source and a garden shed.
- The off-street parking is paved and has exposed brick walls on either side.

The interior of the attached house

- This unit was renovated in 2000 with some modern amenities added while keeping some of the original features.
- The first floor is 656 square feet and the second floor is 577 square feet.
- The house has wood floors on the first floor and carpets on the second floor. The carpets retain contaminants which are harder to clean than wood floors.

Image 3.17.

On the first floor is the dining area and lounge area of the house.

Image by Paul Avery & Shane Bohbrink

Images 3.15, 3.16.

Wolverton, UK, attached house in an industrial town. The long backyard has a deck, garden, shed and off-street parking.

Image by Paul Avery, Shane Bohbrink, and Avenue Estate Agents

Images 3.18, 3.19, 3.20.

Above is the long kitchen with an eating area. The main bedroom is on the second floor. Below to the right is a floor plan of the ground floor and second floor.

Images 3.18, 31.9 by Paul Avery & Shane Bohrbink

Image 3.20 by Paul Avery, Shane Bohrbink, and Avenue Estate Agents

- Double-glazed refurbished sashed windows are on both stories of the house.
- Radiators are used for heating the house.
- There is a gas stove in the kitchen. Gas stoves emit NO_2 (nitrogen dioxide), CO (carbon monoxide), and HCHO (formaldehyde), each of which can exacerbate various respiratory conditions and other health ailments.
- In the kitchen there is a dishwasher, washing machine, and dryer.
- There are doors into each room of the house on the first and second floor.
- There are working fireplaces on the first and second floors. The contaminants from the fireplace include NO_2 (nitrogen dioxide) and PAHs (polycyclic aromatic hydrocarbons)
- There is a ladder to a loft space from the second-floor landing. The loft space is used for storage.

The Green Design Can Improve

Knowing all these factors, the resulting design can improve <u>air circulation</u> with fans on each floor. Planted designs can be created to counteract the contaminants from the <u>gas stove</u>, <u>air</u>, <u>household cleaners</u>, <u>paints</u>, <u>furniture</u>, <u>carpet</u>, and <u>wood sealants</u> on the floor. Additional design solutions can be used for <u>composting</u> and <u>recycling</u>. Additionally, the design can improve the air quality behind the house in the alley where the off-street parking and sheds are located. The alley connects all of the single car parking lots of the connected houses. The series of parking spaces along the alley is divided by brick walls and can be utilized for a plant-based design which filters the car exhaust fumes and leaking car fluids throughout the entire alley. The design applied to one parking space can be applied similarly to all spaces.

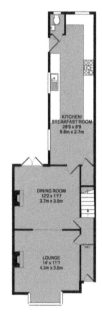

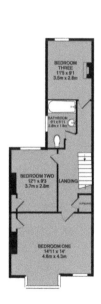

Images 3.21, 3.22, 3.23.

House in Fargo, ND with greening applications. The first-floor main entrance can use green design to improve the air quality from the front entry.

Image 3.22 by Tracy Green/Green Team Realty

*AK: What are some alternatives to **fiberglass insulation?***

SF: There are a few alternatives. Recycled jeans are used by some companies to make insulation. Wool is also an effective insulator while also being fire resistant. There are now soy-based sprays which are used similarly to traditional foam sprays. Further information about these resources can be found at your local and regional building supply company.

*AK: Can the greenwalls contain **soil, hydroponics, and aeroponics in the same piece**?*

SF: This would be a challenging design to create. In part, because the plants for each of these conditions have different nutrient and water needs. Though it is possible to create a design with plants with different water, nutrient, and light needs, it would be difficult. Choosing plants that work symbiotically is most effective for the long-term success of the design.

ENTRANCE HALLWAY AND STAIRWELL

Images 3.24, 3.25.

Plant materials in this house are for improving air quality. A chart of common household contaminants is found in this chapter.

Additionally, given the intense winter season in Fargo, adding interior plants in a design can psychologically improve the homeowner's state of mind. Seasonal affective disorder (further information about SAD can be found online) can be improved with daylight lamps as well as plants which show seasonal changes and flowers for the owners.

Creating some pieces which can add gentle water features and water sounds also adds a calming atmosphere to the home. The quantity of plants shown are for the scale of the contaminants in this house as well as for the psychological benefits.

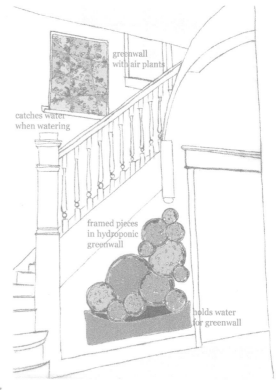

greenwall with air plants

catches water when watering

framed pieces in hydroponic greenwall

holds water for greenwall

*AK: Where do I get **aeroponic plants**? And how do I keep them alive?*

SF: Aeroponic plants can be found in some nurseries or online. They are becoming more and more commonplace and are easy to order. The common term for aeroponics is air plants. Even though they do not need to be sitting in soil or water, aeroponic plants still need water. You can use either a spray bottle with nutrients in the water, or let the plants sit on a plate with nutrient filled water which they can absorb. Most companies that sell aeroponic plants sell nutrients to add to the water.

In aeroponic designs, there are not as many diverse plant choices. Though the plants in aeroponics can be colorful, they are also stiffer. However, in aeroponic designs, the plants need less watering, and there are many creative ways to use these plants. One common use of aeroponic plants on a log in bathrooms. The plants can absorb the constant moisture of the water from the shower while just being placed on the log for aesthetic reasons.

STUDIO DESIGN

Image 3.26.

As any artist will tell you, their studio is the place where the most fun, inspirational, and challenging beauty happens. Creating a design where there are changing colors, changing patterns of growth, fresh air, and pure white light is a beautiful way to inspire an artist. Artists are not the only ones who can have this beauty—everyone can enjoy this design which can also have health benefits.

This design includes plants which improve air quality for those who paint with oil paints and use solvents, turpentine, and mineral spirits. Plants which help remediate these contaminants include:

Fringed Sage (*Artemisia frigida*)

Buffalo Grass (*Bouteloua dactyloides*)

Bermuda Grass (*Cynodon dactylon*)

Canadian Wild Rye (*Elymus canadensis*)

Yellow Sweet Clover (*Melilotus officinalis*)

AK: Are all indoor plants **safe for pets**?

SF: Check the specific plant to make sure it is safe for the animals in your household. Most plants mentioned in this book are safe for animals to sniff or eat, though there are exceptions, and there is continuing research on this topic. Do research the plants you choose on your site, both interior and exterior, to make sure they are safe for your animal friends.

BEDROOM GREEN DESIGN FOR COUPLES TO AID TRANQUILITY

In a study about communication, it was found that people were more open to sharing and helping others when there are refreshing food scents. Further, when sounds of nature such as gentle waterfalls, peaceful birds, and other relaxing sounds are heard, people are again more open to helping and sharing with others. This applies to design and can be used as a design inspiration in bedrooms and other rooms.

IN A BEDROOM, TRANQUIL SOUNDS AND SCENTS CAN ADD PEACE, HARMONY AND ENCOURAGE SHARING

Image 3.27.

Using plants and gentle water sounds in a bedroom design can encourage couples' communication with scents and tranquil sounds. Soothing sounds can be created with water features. Plants which have a pleasant scent include:

Chocolate Cosmos (*Cosmos atrosanguineus*)

Freesia (*Freesia*)

Strawberry Scented Geranium (*Pelargonium 'Lady Scarborough'*)

Lilac (*Syringa vulgaris*)

Herbs such as mint, oregano, basil, rosemary, sage

Tuberose (*Polianthes tuberosa*)

AK: How can I use **_scent in my home_** to help create a healthy atmosphere?

SF: Aromatherapy is the use of scent for healing qualities. Studies from the NYU Langone Medical Center and the University of Maryland Medical Center have shown that inhaling scents has health benefits. These studies specifically conclude that headaches and nausea have subsided with the use of peppermint; menstrual paint has subsided with an abdominal massage using lavender, rose, and clary sage. Anxiety, stress, and depression have subsided with the use of lavender, rose, and frankincense. Pain has subsided with the use of chamomile. Scents can be found at your local and regional natural food shop.

LACE CURTAIN OF AIR PLANTS

Image 3.28.

Weaving together air plants on a lace or open fabric can add a wall of plants as a curtain. The plants are sewn with a fishing line onto a fabric such as a fishnet or other open fabric. Be sure to leave space for the plants to grow and bloom. When the plants need to be watered, gently spray the plants with a water bottle filled with water and plant nutrients. Use a towel under the curtain if appropriate, and clean the window of the water spots afterwards.

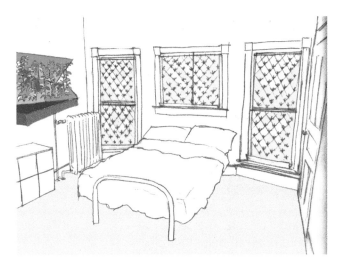

DETAIL OF THE AIR PLANT CURTAIN

Image 3.29.

Air plants are plants which do not to sit in soil or water to grow. Most commonly, air plants are seen in houses in bathrooms on rocks or log designs. In this design, the air plants are gently sewn with fishing line onto the fabric. The fabric is hung like a curtain, and acts as a lace curtain across the window. The lace effect allows light to shine through while blocking views of neighbors or a view from the window.

HOUSEHOLD COMPOSTERS

There are multiple choices when purchasing electric and non-electric interior composting units. However, with some space in a yard an outdoor composter can compost household waste while adding rich nutrients back to the soil. The rich nutrients in the compost can improve the health of the plants in your yard. Properly maintained composting adds health to your site. There are four basic forms of composting.

A low-maintenance pile is suited for people who want something simple, who do not need to use the fertilizer quickly, and who have space in their yard. This is effective for any size household with some yard waste. Simply put your organic yard and kitchen waste into a pile in your yard and let it decompose—no turning of the pile is required. It might take anywhere from six months to two years, but eventually all of that waste will turn into compost. To make your pile more pleasing to the eye, you can enclose it on three sides. About once a year you can dig out the finished compost from the bottom. For faster composting, you can turn the pile over occasionally with a shovel.

Holding bins are suitable for people who want something low-maintenance but more concealed than a pile. These can be used in cold weather. Holding bins come in different scales and styles. With a small household of 1–3 people, 30–40 gallon composters will be fine. Larger bins are best suited for larger households. You can make your own holding bin out of wood. Holding bins offer flexibility in terms of how closely you manage your compost—you can turn your compost for quicker results, but waste will also decompose on its own inside. If your bin has insulated sides, your compost may keep cooking even in winter, though the process will be slower.

Worm composters are best suited for people who want to compost indoors or have small households that do not generate yard waste. The resulting compost can be added to indoor planters. With the use of a 3–10 gallon bucket and a packet of red worms, you can create your own indoor composter. Worm composting, or vermicomposting, is one of the fastest composting methods—each pound of worms will process half a pound of food scraps daily. And it's so compact, you can put your bin under your kitchen sink. Since red worms are so efficient, you don't need to aerate your compost, and your bin won't smell or attract pests. The worms will not process brown waste, meat, dairy, or fatty foods.

Composting services are effective for people who want to compost and are happy to pay a monthly fee for pickup of their compost. Companies pick up five gallon buckets or bins from the curbside weekly. Some communities, such as those in Seattle, Washington require businesses and residences to compost and fine those that put food scraps, compostable paper, yard waste, and recyclables in their garbage.

Images 3.30, 3.31.

Connected house in Wolverton, United Kingdom

The end unit has an exposed brick wall which can be used for a planted wall. Creating a wire structure in front of the wall protects the bricks as it prevents the plants from attaching directly to the house. Using rain water from the roof, the plants on the wired wall can be irrigated mostly by rain water.

DESIGNS FOR GREENWALLS, DRIVEWAYS, WINDOW PLANTERS, AND FRONT GARDENS

Additional plant designs can be used in the driveway with permeable pavers to allow water to filter to the water table. The plants filter the water before it goes to the sewer system and water table.

Exterior window plants utilize rain water from the roof and grey water.

The lined pitched roof is planted with plants which tolerate extreme water and heat changes.

BACKYARD OF INDUSTRIAL HOUSE

Image 3.32.

Permeable pavers are used to help transition from the formal wooden deck to the garden and walkway. The walkway leads to the shed and parking space at the end of the property. Using a transition of tight positioned paving through loosely positioned pavement allows plant materials to grow in abundance throughout the back yard.

BACKYARD SPACES: USE OF STORM WATER AND GREY WATER

The backyard of the house has multiple uses including as an outdoor dining area, garden, walkway, and there is a storage shed and parking space at the end of the property.

Generally, anything that can grow as ground cover in your area which can be walked on, can be used. Based on your location, there are many ground cover plants which can grow within permeable pavers. Effective plants for pavers which can clean car fluids include: Indian Mustard (*Brassica juncea*); Bermuda Grass (*Cynodon dactylon*); Fescues (*Festuca*); Ryegrass (*Lolium multiflorum*); and White Clover (*Trifolium repens*)

Creating a space for a storage unit for the rainwater and grey water is useful for irrigation of the entire backyard. Seasonally, snow is also collected in the storage barrel for irrigation in the spring. When needed, a hose is attached to the storage unit, and basic gravity-fed water pressure is used to water the site. An outdoor 30–40 gallon composting unit is included near the edge of the property for use throughout the year.

AK: How can I use **colors** to improve the different rooms in a house?

SF: Color therapy is the use of colors to affect moods and can be used to improve the overall design of a space. In addition to the deep intense colors of flowers, he differences of greens is also something which can be an asset in the design. After years of living in the American southwest I appreciate even more the differences of grey-greens, blue-greens, lime greens, and pale greens through to intense greens with mixes of violet. Sometimes the subtleties are more visible in different lights, at different times of day, or in different seasons. Enjoy the changes of the colors as they enhance the space.

AK: How do I produce a **tranquil sound** with a water feature?

SF: Water sounds can be tranquil and add a soothing atmosphere to a site. Moving water also aerates the water, adding air to the water while it is in the water tank before it goes through the plants. Both of these are beneficial for the plants and health of the water.

Water sounds are created when there is a gap between the water source and where the water is landing. Consider pouring water out of a glass into another glass. The space between the top water glass and the bottom water glass determines whether the sound is gentle or loud. Additionally, smooth or rough surfaces, and the speed of the flowing water can make the sounds gentler or louder.

In vertical hydroponic plantings where the plants are grown on a capillary pad, creating water sounds is as simple as leaving a gap between the lowest point of capillary pad and the water tank. With a gap of a few inches to a foot or more, you can not only hear the water, but you can also create a visible water feature. As the water may not be continuously running through the green artwork, the sounds will only be audible while the water is running.

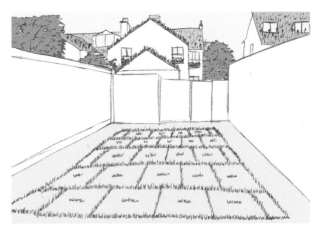

DRIVEWAY, PARKING SPACE

Image 3.33.

Permeable pavers create a surface that combines stonework with plants. Pavers come in a variety of patterns and sizes. All of the designs allow for cars to safely park on the surface while also allowing water to permeate through in the parking space.

This material has a few benefits. First, it allows water to run through the material without creating puddles. Additionally, water is filtered through the plants, allowing the plants to clean the water from car leaks and other surface contaminants.

Additionally, the mixed planted surface slows down storm water before it gets to a city sewer system. Additional information on permeable pavers can be found at your local and regional home building suppliers. In some areas, there are tax rebates for using this material.

AK: How many **plants do I need per square foot** in order to have effective cleaning? Or is it best I measure the space in cubic feet? And what variables do I need to consider in addition to just the square footage?

SF: Spaces with plants have lower concentrations of air pollutants. This is the common trait found in multiple studies over many years. After that, the rest of elements vary and create a complex answer to such a reasonable question.

First, many of the studies were run in laboratories or sealed areas with contaminants being added to the air, and after hours or days with the plants, the level of concentration of the contaminant was reduced. This is great news and was repeated in many studies.

However, the first level of complexity is the size of the plant and the type of plant used compared with the size of the space. This varied greatly throughout the studies. The next level of complexity is the one that makes the question even more challenging—houses and real life situations have a multifaceted combination of contaminants from the users of the space, the materials present, the uses of the space, ventilation, and more. It is this combination that makes the answer specific to each site. There are more and more studies on air contaminants and plants published yearly.

However, with that said, through careful examination of your site, you can begin to answer the question and create a design that works for your site. A study by Stanley Kays measured over 180 different airborne compounds in several houses in Athens, Georgia. In 2010, WHO (World Health Organization) published guidelines which found that indoor air pollutants have concentrations of contaminants which can cause health concerns for the inhabitants. Some of the most common contaminants in indoor air in houses are listed.

KITCHEN

Image 3.34.

This couple loves to cook and use the kitchen frequently, including as a work nook area for one of the men. The greenwalls here use the grey water from the dishwasher and washing machine (located in the kitchen) for their water source. Herbs and edible flowers are in the greenwalls. Additionally, plants for remediating gas stove contaminants include:

Bamboo Palm
(*Chamaedorea seifrizii*)
Red Fescue (*Festuca rubra*)
Tall Fescue (*Festuca arundinacea*)
Soft Rush (*Juncus effuses*)
American Vetch (*Vicia Americana*)

CONTAMINANT	COMMON SOURCES
Benzene	Tobacco smoke, stored solvents, particles from furniture, pressed wood products, nylon carpets, wood-burning stoves, paints, and adhesives.
Carbon Monoxide	Poorly ventilated cooking or heating appliances that burn fossil fuels, tobacco smoke, and burning incense.
Formaldehyde	Combustion emissions, tobacco smoke, pressed wood products, paints, adhesives, cleaning products, and electronic equipment.
Naphthalene	Mothballs, concrete, plasterboard, tanned leathers, paints, insecticides (containing carbaryl), solid block toilet deodorizers, wood smoke, fuel oil, gasoline, unvented kerosene heaters, tobacco smoke, multipurpose solvents, lubricants, herbicides, charcoal lighters, hairsprays, and rubber materials.
Nitrogen Dioxide	Tobacco smoke, stoves, ovens, space and water heaters (especially if poorly maintained), and fireplaces.
Polycyclic aromatic hydrocarbons (PAHs)	Tobacco smoke, fuel stoves, open fireplaces, candles, and incense.
Radon	Stone, soil, and some building materials such as alum, shale, and concrete.
Trichloroethylene	Wood stains, varnishes, finishes, lubricants, adhesives, paint removers, certain cleaning products, and drinking water in contaminated areas.
Tetrachloroethylene	Adhesives, fragrances, spot removers, stain removers, fabric finishes, water repellents, wood cleaners, motor-vehicle cleaners, dry-cleaned fabrics, and contaminated drinking water.

Once you have looked around your house and made a list of possible sources of contaminants, and determined how they affect your air quality the next step is determine what you wish to do. For example, if you have looked at your cleaners and realized that there are many contaminants in them you may choose to either switch the cleaner or keep the cleaner and use a design solution to reduce the air contaminant. Or both.

The use of plants for cleaning the air, soil, or water is called phytoremediation. There is a database link at www.engaginggreen.com as well as other sources online and green professionals who can help determine which plants may be most effective. When doing your own research, you will likely find a variety of answers out there, as studies are continually being completed and there are many different areas to study.

Once you determine which plants are most effective, consider using the plants in combination so that the contaminants are reduced by a variety of means. Use as large a plant as you can comfortably maintain that works with the space. Some studies have shown one large plant per 100 square feet can be effective while others show three plants per office are effective. Other studies vary based on the plant size and choice, and say that the ratio of plant to space is 10-25% of the cubic space. There are great variances, with many similar studies showing that indoor plants improve mood, increase performance of creative tasks, accelerate recovery in hospital patients, and that caring for plants makes people happy.

ENTRY OF THE HOUSE

Image 3.35.

The plants used in this design include those which help clean the common contaminants noted earlier. A more complete search can be done by using the phytoremediation database at www.engaginggreen.com

PLANTS IN THIS HOUSE INCLUDE

Chrysanthemum (*Chrysanthemum x morifolium*); Orchid (*Dendrobium taurinum*); Buffalo Grass (*Bouteloua dactyloides*); Canadian Wild Rye (*Elymus Canadensis*); and Red Fescue (*Festuca rubra*)

AK: *How can I **get my air tested** in my home, and **whom can I contact** to review my home if I have questions?*

SF: There are professional companies which can test the air in your home as well as mail-in tests. Recently, room air quality monitors have been created which can be placed in the room to monitor, measure, and give reports of specific air contaminants in the room. These products are being updated yearly. All of these different choices vary in price and vary in which contaminants they measure. Choosing the option that works for you, as well as finding a professional to help with your design needs is always an option.

BEDROOM OF CONNECTED HOUSE

Image 3.36.

Carpets contain chemicals from cleaning products, pesticides, and carpet cleaners, which release harmful substances into the air and degrade air quality.

The continued plant list for this house includes:

Gerbera Daisy (*Gerbera jamesonii*)
Ryegrass (*Lolium multiflorum*)
Empire Bird's-foot Trefoil (*Lotus corniculatus*)
Alfalfa (*Medicago sativa*)
White Clover (*Trifolium repens*)

Images 3.37, 3.38.

Painted snow and painted yard Fargo, North Dakota and sculpted snow in Winnipeg, Canada

Temporary designs to show color, form, and change through different seasons.

Image 3.37 by David Samson/Forum News Service

PAINTING AND SCULPTING SNOW

Working in the context of where you live and its conditions is important. In Fargo, ND the winter season is legendary. Working with the snow as an asset for design is beneficial for enjoying the season. Painting the snow with bright colors, patterns, artwork, flowers, or a poem provides the emotional benefit of seeing color in the winter months of continued snow.

Upon each snowfall you can change the design using non-toxic paints such as silk paints in dozens of rich deep colors. Using a three-gallon or five-gallon weed sprayer is a fun and easy way to cover larger areas of space with color.

*AK: How are **temporary designs** beneficial?*

SF: Short-term and long-term designs have different roles. Short-term temporary designs can take the role of continuous change with color and form. This also allows for temporary colors and designs during a season when views or landscapes may be bland or repetitive. Painting landscapes allows a short-term solution while the long-term designs are still in process. The psychological effects of seeing change and color in a landscape have been shown to be beneficial in studies

*AK: Does **painting living trees** harm or kill the tree or the bark?*

SF: That is based on the paint you use. Do not use latex paint, house paint, acrylic paint, oil paint, or other common paints. These paints are harmful to trees and animals. Paints which are used on food, are safe for children, or are completely washable with water are safe for painting trees and allow trees to breathe and animals to use the trees naturally.

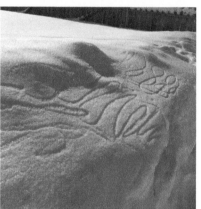

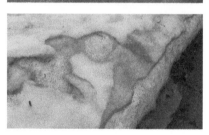

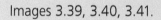

Images 3.42, 3.43.

Sculpted snow works can be anything from a series of small sculptures through to large pieces and installations over 30 feet tall and wide. Sculpting snow works best with a packed block of snow, from which you can carve out shapes and forms. Let the snow sit in the block form for at least 24 hours during cold weather before starting to sculpt for best results. Do not add water—carving ice is more challenging and requires tools not typically found in a household whereas carving snow can be completed with hand tools.

Images 3.39, 3.40, 3.41.

As the snow melts throughout the season, the layers of the colors are a spectacle to view in unexpected and unplanned reveals.

*AK: Where can I **purchase paints** for temporary land-scape designs which are safe for the plants and site?*

SF: Sources for food paints, water-color paints, silk paints, and chil-dren's chalk paints include local food, culinary, art or craft stores, as well as online stores. The paint can be applied with hands, brushes, spray bottles, or a 3 or 5 gallon weed sprayer for larger areas.

PAINTING AND SCULPTING GRASSES AND YARDS

Images 3.44, 3.45, 3.46.

Painting a site can be done in any season with water-based non-toxic paints. This can be applied to the grasses, branches, outdoor furniture, house or garage. There are multiple psychological benefits of having colors in your home. Temporary designs allow for a quick change, the fun of color, and experimentation. The paint options include silk paints, chalk paints, food coloring, food paints, and watercolors applied with a weed sprayer or brushes. With either rain or a watering hose, the design is easily removed and a new design can be created.

A FEW NOTES AS YOU ENJOY YOUR GREENING PROCESS

Your home is where comfort, joy, and health are the main design components of turning a house into a home. Creating an environment which has emotional and physical health through designs which grow, change, and develop throughout time is a healthy pursuit. As you look through the opportunities that are specific to your space and your changing family, mindset, or technology, you will find new ways to improve your home environment. Use research and technology along with scent, color, form, function, and aesthetically beautiful design to create a home of your dreams. Dream big, and live it out loud.

Chapter 4

Healthcare Facilities

INTRODUCTION

Green Solutions for Healthcare Facilities

The combination of physical, mental, and spiritual health are the building blocks for a fulfilling healthy life. This chapter looks at the psychological benefits of adding green design to a variety of healthcare, daycare, and supportive living facilities to benefit children and adults.

There are a vast range of healthcare facilities which range in scale, needs, whom they care for, and the average amount of time people spend there. Below is a very brief list of healthcare establishments which would benefit greatly from a planted design.

The brief list of healthcare facilities includes:

Public Health Care Facilities
Adult day care services for adults with disabilities and special needs
Assisted living facilities
Walk-in clinics
Private doctors' offices
Medical offices with multiple practitioners
Hospitals
Specialties such as
    Pediatrics
    Obstetrics & Gynecology and Women's Health
    Psychology
    Psychiatry
    Dentistry
    Optometry
    Cancer treatment
    Physical therapy
    Acupuncture
    Chiropractor
    Massage therapy

These types of facilities share similar elements such as:

Parking lots
Entrances
Hallways
Common rooms or waiting rooms
Private offices

In these areas, unique gardens, greening and healing designs can be utilized to improve the space for all users.

Existing Site Conditions

Let's face it, being sick or in pain is not a feeling we would choose for anyone. Most people would rather choose joy, laughter, good health, vibrancy, and other feelings of optimal health. Therefore, when we do not feel well, whether it is emotionally or physically, getting help from others in a positive and supportive environment helps ease the return to optimal health.

People who are exposed to green planted designs and can see them while recovering have been shown to have lower stress, lower blood pressure, and it also improves the pace of healing and reduces hospitalization time. The book *Healing Gardens* by Cooper Marcus and Barnes, discusses the design philosophy of healing gardens. One study of an outdoor space in hospitals showed that 95% of the people surveyed had positive mood changes after spending time outside. When participants were asked which specific elements changed their moods, more than 66% said trees, flowers, colors, seasonal changes, and greenery had a positive impact. Fifty percent of the participants also mentioned that sensory details such as sound, scent, and touch also had an effect. This included elements such as fresh air, birds chirping, and the sound of water (such as fountains).

The benefits of green design are not only valuable for the patients, but also for the practitioners, administrators, and all users of the site. Medical practitioners and administrators who care for the patients are also in need of taking care of themselves. One cannot care best for another unless they take care of their own personal health, mentally and physically. Also, people who attend appointments with patients are likely to be stressed or in need of healing. Everyone can benefit from more self-care and joy in their own lives. Green spaces help improve focus, clarity and healing.

Gardens at hospitals have been shown not only to have patients and their families and friends attending, but also doctors and administrators. Selecting plants which bloom throughout the year allows for cycles of colors, scents, and views, and illustrates the passage of time, seasonal change and growth through the garden design. Everyone who experiences a successful year-round blooming site benefits from the healing properties of plants.

As the scale and details of healthcare facilities vary from site to site, creating designs which are flexible, perhaps mobile, and always with plants which can grow and change, can help meet the site's needs. When designing for a healthcare setting, the more greenery, the better. This is not only for the improvement of mental health, but also for air quality. Though healthcare facilities typically have an engineered system to ensure air quality, green designs help support the engineered system.

Plants, water features, textures, and tactile experiences soothe the spirit. The sounds and scents of plants, water, wind, and wind chimes create a mood, a cadence, a spirit of calm and peace.

WHAT TO DO NEXT

Questions to Ask about Your Site

It is now time to look at your site and to enjoy the healing properties of plants. Time to dream big. Looking at the sensory experience of designs is the next step. There are multiple senses to use in design—sight, sound, touch, smell, and taste. It is the combination of senses that helps makes the experience of design more memorable and effective. For these designs, four of our five senses are considered, skipping taste.

Scent has close links to memory. A scent from 20 years earlier can bring back a memory long forgotten. Some examples of a memorable scent may include a grandparent's perfume, a house fire, or a meal made by a relative. The memory connected with the scent may be in the back of your brain for an extended time period. However, 20 years or more later, walking by a house recently on fire, a person with the same scent as your relative, or the same meal, can bring the memory quickly to the front of your brain including with all the feelings connected with it. There are studies being published yearly which illustrate the relationship between scent and memory.

Scent can be used to consciously recall memories. For instance, if one were to study with a particular scent present throughout a study session, exposure to the same scent at the time of a test can help the person to recall what was studied during the earlier session. For example, using a scent that is mainly associated with studying, such as cinnamon, can help bring the memory back to the test taker if a cinnamon stick is brought into the testing room.

Using scent when designing with plants is a way to help people either focus or relax. Scent can bring a smile to your face easily, bring back a great memory, or create new ones.

Tactile experiences in design can include touching a soft plant such as Lamb's Ear, or running your hands over a smooth river stone, or feeling a perfectly fitted chair against your skin. Physically having an object against your skin brings another sense into your experience of a space and helps people connect more with the experience. Tactile experiences, such as petting a soft cat, can also relax a person. An element of touch can be created with soft materials for comfort, aged wood with worn marks for implications of a weathered past, or cool steel for clean cool sleek smoothness.

Sound features are a fantastic way to subtly or dramatically create atmosphere. Consider soft gentle music in a space versus loud fast paced heavy bass music. Similarly, water features can be subtle or dramatic with tranquil soft sounds or a crashing waterfall. It also may be possible for the same water feature to have multiple settings for different effects.

 Sight has many avenues to address—color, scale, foreground, middle-ground, background, textures, shapes, forms, etc. The psychological effects of color have been studied by artists and designers for many years. Consider what colors you enjoy and what moods they bring about for the users of the facility.

Specifically, medical facilities lean towards lighter colors such as pastels to help relax patients. These colors have been found to be more effective in reducing stress and promoting relaxation. This is important to know when designing with blooming plants in healthcare facilities.

Knowing all of these factors, look for possible spaces where visitors, administrators, and professionals are situated throughout the working hours of the facility. Even if your site does not have large open horizontal or vertical spaces, having multiple smaller pieces throughout the space with three common features can be quite effective. The common elements can be the materials, color, scale, texture, or scent. Common features draw a person's eyes through a space and connect the space. Using this design technique with a green design can help the space to feel fresh and create a healthy site design.

Additionally, using plants which also remediate the harmful effects of the materials on the site, as mentioned in other chapters, is also beneficial. Healthcare facilities typically use cleaners with chemicals, have furniture with fabrics which likely contain chemicals already, have freshly painted areas with paint VOCs (the elements which causes paint fumes), wood sealants, carpets with VOCs, and other common contaminants. One way to help with carpet VOCs is to air out carpets for 30 days before occupancy. This allows many of the contaminants to be released into the air and cleared out.

However, while other chapters look more specifically at physical materials which plants can be used to remediate for the effects of, this chapter looks at the emotional and psychological effects of green design. Please read through other chapters for more information on plants for remediation of contaminants in the air.

The two examples of greening applied to healthcare facilities include an assisted living facility and a doctor's office that specializes in pediatrics. The green ideas illustrated in these two facilities can be applied to many types of healthcare sites, including those listed earlier. The scale of the facility and users will determine the most effective design. Ashley Kaisershot (AK) asks questions of the author, Stevie Famulari (SF) for more insight and clarity.

Example A—Retirement and Assisted Living Facility

The property and common areas of the facility...

- There is 15,000 square feet of living space for the residents.
- The exterior patios and gazebos are screened for protection from insects.
- There is a parking area adjacent to the building.
- In the adjacent park there are walking trails which are accessible for a range of levels.
- The adjacent park also has a variety of trees and plant materials.
- There are interior and exterior common areas available for use by the residents, family and staff.
- The interior common spaces include a recreation room, library, family visiting room and television room.
- There are both carpet and wood floors throughout the building.
- The common areas are used by the staff, residents, and guests, who vary greatly in age and physical abilities.

Images 4.1, 4.2, 4.3.

The assisted living facility seen here is a compilation of typical facilities for short- and long-term living. Common areas include an outdoor garden, walking trails, indoor recreational rooms, and dining areas.

Images 4.1, 4.2 by iStock

Image 4.3 by MarcusPhoto1/Getty Images

Private and shared rooms...

- The private rooms range from 350 through to 450 square feet.
- The shared rooms range from 450 through to 650 square feet.
- The rooms have large windows with a view to the exterior.
- Artificial lighting does not currently use daylight spectrum bulbs.
- Window air conditioner units are in each private and shared room.
- The large bathrooms allow space for those who use walking assistance equipment. The showers have no barriers, allowing for walker and wheelchair access.
- Housekeeping and laundry services are used for both common and private spaces.

Hallways…

- The hallways are wide enough for wheelchair access, including the ability to turn the wheelchair around.
- The hallways have artificial lighting which are not full spectrum bulbs.
- There is seating and artwork in the hallway.
- The hallway connects the residents' private and shared rooms, administrative offices, and common rooms.
- The hallways act as a shared public space, but also the entrance to the private and shared residents' rooms. This area acts as the porch to a house, and can be considered from a design perspective as the entry point to rooms.
- Sounds echo in the hallway because of the hard surfaces.

Administrative offices…

- Administrative offices range in size from approximately 225 square feet to 400 square feet.
- An employee lounge and conference room is also located on-site.
- Some administrative offices have a window to the exterior, some no windows.
- The offices have furniture made of wood, fabrics, and metals. The floors of the offices are artificial wood and industrial carpet.
- Air circulation in the offices is through the engineered ventilation system.
- Lighting in the offices is not full spectrum bulbs.

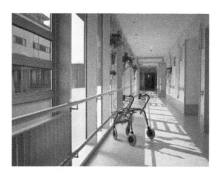

Images 4.4, 4.5.

The hallways of the facility connect the private rooms, shared rooms, administrative offices, and common rooms.

The private rooms have a bed, closest, dresser, television, and bathroom.

Image 4.4 by Westend61/Getty Images

Image 4.5 by iStock

The Green Design Can Improve

Knowing all these factors, the resulting design can lower blood pressure and stress in people who see or work with the plant materials. Focus and clarity can be improved using plants with scents associated with memory. The energy of a space is invigorated with plants which have different stages of growth and bloom for an extended period of time. Interior plants may require full spectrum bulbs, which also improve eyesight. Working with plants can improve physical health by caring for and tending to the plants as they grow. Additional design solutions can be used for composting and recycling.

Example B—Pediatric Office

These images are a compilation of multiple different pediatric offices with commonly shared room types.

A medical care facility building

- The drive into the parking lot is the primary entry for this building.
- The drop-off area in front of the building allows people to drop others off close to the entry.
- The walkway in front of the building is a concrete sidewalk with severely pruned shrubs planted on each side of the door.
- Plant materials are kept to a minimum on each side of the doorway.
- The building has a pitched roof and many windows on all sides.
- The parking lot of a doctor's office holds a range of vehicles from 15 in urban areas, through to 40 in larger offices.
- There is night-time lighting of the front of the building in the shrubs and at the door entry.

Images 4.6, 4.7.

The entry to the building includes a parking lot, walkway, and sidewalk prior to entering the building.

Image 4.6 by iStock

Image 4.7 by ESB Professional/Shutterstock.com

The spaces within the facility

- The interior entrance hallway of the office has a closet for coats and umbrellas, as well as holding additional games and toys not currently in use in the waiting area.
- The waiting room of a doctor's office holds a range of people from 15 through to 30. There is typically more space in a pediatrician's office as most patients come with one or more family members.
- The front reception area includes space for one or more people to greet the patients. Computers, files, desks and artwork are typical in this area.
- Private administrative offices to discuss finance and other private needs range in size from 200 square feet to 350 square feet. The majority of the space is for patients and equipment.

- The hallway leads to private medical offices, administrative offices, and the waiting room. The floors in pediatric offices are typically not carpeted.
- The second story has a private lounge for the employees.
- The private medical offices have a range of equipment and are usually lit with artificial light which is not full spectrum. For the privacy of the patient, there are typically no windows in medical rooms.

Image 4.8.

The waiting area can be shared by people of a variety of ages. Some people may be waiting in this area while a parent goes into the private office with the child.

Image 4.8 by Altrendo Images/ Getty Images

The Green Design Can Improve

Knowing all these factors, the resulting design can reduce stress in those who work and visit the office through a green design which starts at the exterior of the site and continues throughout the interior. The design solution can include planter strips in the parking lot, planted sidewalks, a garden on the building roof, and a planted entry. The exterior plants can use rainwater for irrigation. Inside the building, where clients and employees spend more time, greenwalls and planted designs in the common area and private rooms can help to lower blood pressure and promote healing throughout the site. Grey water from the sinks can be used for irrigation of the interior plants.

Creating a healthy space for short-term visitors as well as daily employees can ease some of the challenges associated with the healthcare site.

Images 4.9, 4.10.

It is common to have a waiting room and private medical offices in a doctor's office. Though the style may vary from one office to another, chairs and play areas are common in the waiting area. A medical table, medical supplies, chair, and a computer station are common in a private office.

Image 4.9 by iStock

Image 4.10 by cassiohabib/ Shutterstock.com

Image 4.11.

Assisted living facility in Florida with greening applications.

The common room of the living facility is where people spend time with family and friends.

COMMON ROOM OF AN ASSISTED LIVING FACILITY

Image 4.12.

The design of the waiting area creates an atmosphere of peace and tranquility through details such as scented oils, relaxing colors, water features with tranquil sounds, daylight bulbs, and a small Zen-style sand area for clients to release their anxiety.

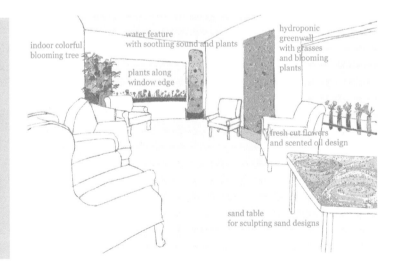

indoor colorful blooming tree

water feature with soothing sound and plants

plants along window edge

hydroponic greenwall with grasses and blooming plants

fresh cut flowers and scented oil design

sand table for sculpting sand designs

ENTRANCE TO THE BUILDING

Image 4.13.

The entrance is an opportunity to set a mood for the space. Most building entrances are also their exits, creating the last view a person sees of a site. Creating an entrance that relaxes the client with colors, scents, textures, forms, shapes, and sounds is a great design opportunity for any site.

Entrance design is not only for the visitors, but also for the every-day user—the employees and people who live on site. The frequent users of the site can enjoy the daily change of the blooming plants, growing vines, and seasonal changes of leaves.

An entrance can be both exterior and interior. In this case, the entrance includes the parking lot and the doorway entrance to the building. As a person gets out of their vehicle and walks towards the door, a path of greenery leads the users directly to the front door.

Using scent at the entrance subtly creates an atmosphere that can relax the clients, caretakers, families and all who visit. Using a series of plants which bloom throughout the year also creates a welcoming sight.

AK: What are some of the **techniques of design** that would be helpful for a non-designer to learn?

SF: There are many techniques and many different types of design which we could discuss. Some of the ones which I use when looking at a site include looking at the spirit, the stories, and the needs of the site. This means looking not just at the functional elements of lighting, distances, and weight of the structure, etc., but rather looking beyond those to the soul of the site. This means looking for the story, the flow of people, the environmental possibilities, the healing of the site through plant materials, the ways in which the people here could be healthier and other forms of health and healing. Looking for how to bring those feelings through in an applied design with materials, scale, form, texture, color, scent, etc. is one of my approaches.

I also talk with the users, understand who they are, and why they are on a site. Understanding who they are and what they value helps me to create a design. I also research, meditate, talk with people, and draw out many ideas in loose sketches. The loose sketches put a shape to the words said by the users and allow a visual for people to understand the design more clearly.

Consulting a professional or team of professionals is always an option for an effective collaborative design. Working with people's needs, means listening, then educating people about the possibilities of using design to create a healthy space.

PRIVATE OFFICES

Image 4.14.

In a small private office there is typically not a large amount of space to be creative with design. With the required furniture and supplies, being creative in an office can be useful for green design.

First, consider using a planted greenwall where you may typically place artwork. Many offices have a framed photograph or painting of a feel-good scene. In dentist's offices, sometimes this is even on the ceiling. Consider putting plants in a framed design, including vines along the ceiling.

*AK: Do the designs in this chapter apply to most healthcare facilities? What about **other types of sites that are not for healthcare**?*

SF: That is a fantastic question. The designs throughout this book are for inspiration, education, and instruction for a variety of sites. There are features in each chapter which can be applied to an assortment of sites. For example, the information about plants for remediation of the effects of nicotine may be useful for sites other than apartments, such as entrances to buildings. People sometimes smoke a cigarette nearby building entrances. Using plants to help clean the air as the door opens and the contaminants enter the building may be useful for a variety of sites. Though you may start off looking at the chapter for healthcare facilities, I recommend looking through the entire book for elements and ideas which may be applied from different chapters.

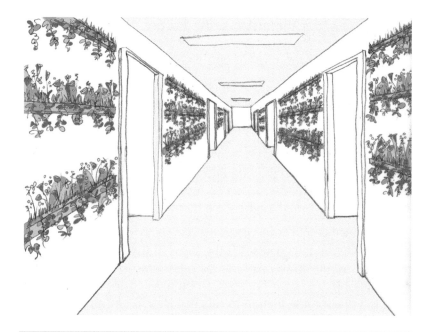

HALLWAY OF THE ASSISTED LIVING FACILITY

Image 4.15.

Whether the hallway is visible from the common room or all the private suites, the hallway continues the green design which was started at the exterior entrance, continues in the common rooms, and then through the hallway. The private suites use some of the same materials, colors, and textures to connect the whole site together.

The hallway space uses shallow planters which are high on the wall and have hanging plants. The foliage from the hanging plants are woven onto the wired form along the wall in an artistic design.

AK: Will the __germs and bacteria__ in healthcare facilities effect the plants or the design solution?

SF: Plants are able to live even on sites with germs and in a great variety of sites. The sites mentioned here are more than suitable for plants to grow. Addressing the lighting and other basic needs of the plant is what will help them to grow well.

AK: Will the __chemical cleaners__ in medical offices kill the plants in the rooms?

SF: Cleaners with chemicals should not be sprayed directly onto plants as complex chemicals can be harmful. Cleaners for medical offices is a complex topic as these types of offices have different standards by which they must adhere to. Because of these standards, the cleaners of choice for some offices use harsh chemicals. Though there are natural alternatives discussed in a different chapter, these alternatives may not be suitable for all sites.

One option is to use natural cleaners if possible, and chemical cleaners if needed. Spraying either cleaner on a plant is not beneficial for it. When needed, spray the area around the planted design, or move the planted design if appropriate to complete cleaning the medical facility.

AK: How are __Zen-inspired design__ ideas relaxing or stress reducing?

SF: In their traditional design, Zen gardens are minimalist in their materials and form. The Zen space or Zen object is used by a person to feel time slow down and their breathing become more relaxed.

In a medical facility, designing a piece which uses sand or small stones in a small contained space is the focus of the design. The sand can be moved with an object such as a stick or miniature rake, to create patterns in the sand. The process of moving the sand, either with the object or your fingers, can be relaxing for a person. The smooth tactile experience of soft sand is a pleasurable texture to touch. Creating patterns in sand allows a person to focus on the texture of the sand, the patterns of the sand, and to sort through and heal the thoughts in their head.

AK: Are there **plants** that people are **commonly allergic** to?

SF: There are plants which produce pollen which become airborne. Some plants produce more pollen than others, which when released into the air cause sneezing and itching for some people. With a wide variety of plants to choose from, there are only a few that produce large amounts of pollen and are common for allergies. A list is included of the plants people are most commonly allergic to. Sources of common high pollen plants can be found at your local nursery.

A LIVING DESIGN FOR THE SUITE

Image 4.16.

A living suite for one or two people can include planted columns inside the room, as well as planted wall pieces. Additionally, making sure the views from the windows have plant materials can support a healthy lifestyle.

Even if the building is near another structure or does not have a view of a park, putting planters outside the window is an option for most sites.

Plants both inside the suite, as well as on the outside, create a growing, blooming viewpoint for residents.

Flowers/Herbs
Amaranth (pigweed), Chamomile, Chrysanthemums, Dahlias, Daisies, Goldenrod, Ordinary sunflowers

Shrubs/Vines
Cypress, Jasmine vine, Juniper, Wisteria

Trees
Alder, Arizona Cypress, Ash (male), Aspen (male), Beech, Birch, Box and Mountain Elder (male), Cedar (mountain cedar, (male), Cottonwood (male), Elm, Hickory, Juniper, Red and Silver Maples (male), Mulberry (male), Oak, Olive, Palm (Date Palm and Phoenix Palm) (male), Pecan, Pine, Poplar (male), Sycamore, Walnut, Willow (male)

Grasses
Bermuda, Fescue, Johnson, June, Kentucky Bluegrass, Orchard, Perennial Rye, Redtop, Salt Grass, Sweet Vernal, Timothy

Weeds
Cocklebur, English Plantain, Lamb's Quarters, Ragweed, Russian Thistle (Tumbleweed), Sagebrush

Image 4.17.

Pediatric office common area. The common area can use a scale which is friendly for toddlers, young children, teenagers, and adults.

COMMON AREA INTERIOR GARDEN

Image 4.18.

The common area has items which are attached to the walls permanently, as well as mobile elements such as green vertical walls on wheels and horizontal planters which define spaces. The combination of horizontal and vertical elements is used to design a space rather than merely objects in a space.

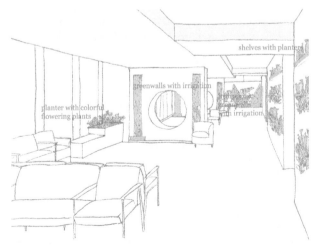

OUTDOOR ENTRANCE TO THE PEDIATRIC OFFICE

Image 4.19.

In any setting, the entrance to the site becomes a unique space with a parking lot, sidewalk and building front. Designing an entrance with planter strips in the parking lot, sidewalk plantings and outdoor greenwalls makes a powerful statement before the client even walks through the door.

Outdoor green design in an urban area becomes a balance between making space for the large number of users of the site, and making the space healthy for all users. More specifics on urban design are addressed in chapter five.

In this design for the office entry, water is captured from the roof and used to irrigate the plants on the building façade. Grey water from the building can be used to supplement the rainwater as needed. The plants on the building façade are growing along a wire which is attached to the exterior walls. The vines growing on the wire are maintained to stay solely on the wired area, so as not to cover signage or windows.

Vines for this design are chosen for their ability to grow in the area, their seasonal color changes, and their fragrant scent. Other plants are chosen for their remediation of urban air.

*AK: Do **custom greenwalls** need to be made for my site or can I purchase **prefabricated greenwalls**?*

SF: Both are great options which can be done in combination. If you find a prefabricated design that works for your site from a nursery or online, by all means, use it. If your site is in need of specific materials, or a specific size, shape, or form that makes it more suitable to have a custom piece, then that is also a fantastic choice. Choosing what works for the site, your budget, your time frame, and your needs is what is important. Greening on small scale through large scale is a healthy choice. Starting small while planning for expansion is one option. I recommend creating a master plan for the expansion with the full scope of the design so you know the direction you are going. As you find pieces which work for the design you can move forward with different phases as you are ready.

*AK: Is **rain water safe to use for plant irrigation** in urban and other areas? You mention air contaminants. If they are in the air, then they must also be in the water that is on the roof, and is then being used for plants?*

SF: The rainwater is safe to use for the irrigation of plants. Some of the contaminants in the air do eventually make it to the surface of area—whether that be a roof, sidewalk, bench, or other. When it rains, the water will run across the surfaces and wash away the contaminants. The water can then be used for plants because the plants can not only stay healthy while using that water, but can also clean that water while it filters through the plant.

There are different levels of water quality, from the cleanest levels of drinking water, through to less clean water which may be used for washing, swimming, fishing, irrigation, or water features. Plants are a natural filter of water, and can live healthily when irrigated by water that is below drinking water standards.

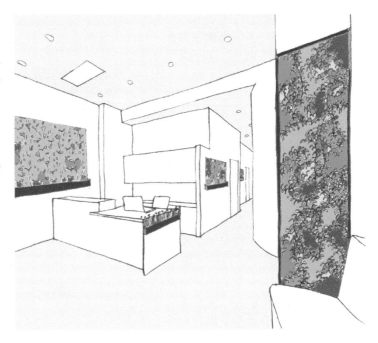

ADMINISTRATIVE AREA

Image 4.20.

The office walls are used for a series of small green artworks throughout the office. Daylight bulbs are focused on each of the planters, while the plants are put on a wired design across the walls.

Though the wall pieces may vary in size, common elements such as the material of the structure, color of the blooming plants, or the shape of a feature can tie all the wall pieces in a space together.

PRIVATE MEDICAL ROOM

Image 4.21.

Similar to the administrative area, the small room's light is improved with daylight bulbs, and the indoor vertical garden allows for plant materials. Plants soothe the soul, which is useful for children and adults in medical facilities.

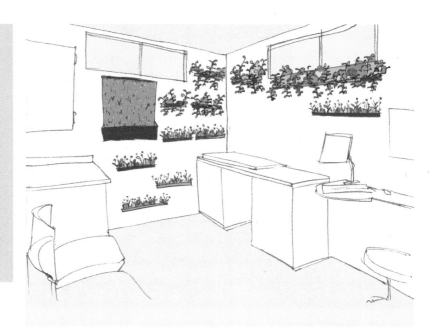

CEILING DESIGN IN A PRIVATE MEDICAL OFFICE

Image 4.22.

In this design, the use of the ceiling for hanging plants and vines creates a site for the patient who may be lying on the medical table while waiting or being treated. The ceiling design includes vines and air plants for ease of maintenance.

CAFÉ WITH PLANTED HERBAL WALL

Image 4.23.

This indoor café area has vertical herbs planted along the windows for use in the food, as well as for a fresh scent while sitting and eating.

Watered as scheduled, this vertical design allows people to sit in the space and interact with the plants while exploring the connections between food and modern design.

Image by ESStock/ Shutterstock.com

CAMPUS STUDY AREA

Image 4.24.

Research at Cornell University's Human Ecology Department found that nature can help people reduce stress. During finals week some professors decided to apply their research to a temporary design outside the Mann Library.

Image by Craig Crammer, School of Integrative Plant Science, College of Agriculture and Life Sciences, Cornell University

PLANTED INTERIOR TREE DESIGN CONCEPT

Image 4.25.

The tree in the center of the space defines the space. Modern and traditional spaces can allow the tree to live and thrive within the room itself.

This is an illustration of a concept. This idea can be used to not only create a design to keep an existing tree within a site, but also to grow new ones to create long-term interior designs.

Image by Zastolskiy Victor/Shutterstock

LIVING TABLE PLANTER

Image 4.26.

With integrated water reservoirs, furniture can help to create an interior garden that soothes the soul. Design companies have prefabricated designs and modular designs. Custom design pieces for spaces are also available by designers.

Image by David Brenner & Habitat Horticulture

AK: Why do you suggest to put a plant with a relaxing scent specifically at the entrance to a medical facility?

SF: The entrance to a site is a design opportunity to set a tone and mood for the space. With scent being notably connected to memory, adding a plant that has an enjoyable and memorable scent at the entry can subtly yet profoundly create an atmosphere. As a person enters the site, the scent of the plant can set a tone of relaxation with lavender, or awareness with peppermint, or food pleasure with mock chocolate, or a great variety of plants. When you begin to associate the pleasant scent with the site, it begins to create a mood and atmosphere. The plants may be either exterior or interior in the design technique. It is more important that they are at a main entrance.

Some plants with fragrances are below. There are many more to choose from. Have a scent test: go and smell some of these for yourself and choose plants that make you and others on the site feel good.

Plants with a pleasant scent:

Basil
Catmint
Daylily, many varieties
Garden Phlox
Gardenia
Heliotrope
Hyacinth
Icternia Salvia
Lavender
Lemon Balm
Lilac
Lily of the Valley
Nicotiana
Peony, many varieties
Rhododendron
Rose, many varieties
Rosemary
Sweet Alyssum

INTERIOR CHILDREN'S GARDEN

Image 4.27.

At the Longwood Gardens in Pennsylvania, one interior garden is designed on a child's scale. Created in 1987 and still enjoyed today, the space includes hands-on water features, sculptures, secret stairways, and plants blooming in different seasons.

Image by Candie Ward for Longwood Gardens

LIVING OFFICE WALL

Image 4.28.

This living wall is created with plants grown in a hydroponic system. The pots are attached to a wall and set on a timed system to water the plants.

Succulent plants which require less water can also be used in living walls. Succulents store water in their tissue, allowing them to grow in dry conditions, and thus needing less water than other plants. Most succulents are easy to care for, grow, bloom, and are happy indoors with direct or sometime indirect light.

Image by iStock

A FEW NOTES AS YOU ENJOY YOUR GREENING PROCESS

This chapter focuses on the psychological benefits of greening as an important element of design. Using the information from this chapter in combination with the knowledge from other chapters about remediation and other design practices can help you to create a design which improves your mental and physical health. With the many types of health facilities which are present, using all possible resources to improve the health of clients and professionals is an advantage for everyone. This is a beneficial approach for sites such as medical facilities, where people may need some extra care, comfort, and healing.

Chapter 5

Buildings in Areas near Airports, Highways, and Train Lines

INTRODUCTION

Green Solutions for Buildings near Airports, Highways, and Train Lines

People who work or live near airports, expressways, highways, freeways, train lines, and subways may feel the vibrations, hear the loud sounds, see the routine traffic patterns, and breathe in the contaminants of the air. Because of the close proximity residents have to these sites with a high number of contaminants, improving the health on a room scale, building scale, and community scale can help to counteract the amount of pollutants that residents are exposed to.

Neighbors of airports, expressways, highways, freeways, train lines, and subways live with the effects of these areas. These include loud noises, lights, vibrations from the movement of trucks and industrial machinery movement, hazards caused by the moving cars on the nearby roadways, the industrial views from one's home, and declining air quality caused by contaminants of the planes, trucks, automobiles, trains, and subways.

People in urban areas as well as those in rural areas may feel the effects of the airport noise and air quality for miles around the site or route of the airplane take-off or landing. As an example, www.statisticatlas.com states that the amount of people who live within a five mile radius of LaGuardia airport was reported as over 2.2 million people in 2015 and that within a ten mile radius there were approximately 7.4 million people. The quantity of people that used the airport for travel in 2016 was 129.8 million passengers according to *"Port Authority Airports set new Record with more than 129 Million Passengers Traveling in 2016" (Press release, The Port Authority of New York & New Jersey).*

The mental effects of living in these areas may be underestimated, with decreased air quality due to contaminants, as well as constant noises and lights, all of which can have short- and long-term effects. These effects may range from sleep disruption in animals, children, and adults as well as long term decline in physical health and mental focus.

The examples in this chapter are of a public school and a residential home. The public school in the Bronx, New York is within five miles of the airport. The house site in Little Falls, MN is within a quarter mile of the freeway, less than 1,000 feet.

Choosing only two sites for this closing chapter was a challenge because there are numerous sites to choose from in this modern culture. To not live near airports, expressways, highways, freeways, train lines, subways, or other conditions of modern culture is uncommon. The severity of these conditions and the ease of improvement may vary. I am not advocating moving out of urban areas or areas near airports, highways, and train lines as the convenience and enjoyment of living in these areas is a preference for many people. I am advocating for the area to be improved with green design solutions to help mitigate the conditions of airports, highways, and train lines.

There are many sites which have conditions which can be made healthier by green practices and design on large and small scales. The green solutions in this book explore healthier designs for sound levels, views, air quality, water quality, lighting, psychological enrichment, and other conditions of sites.

Existing Site Conditions

In sites which are located near airports, expressways, highways, freeways, train lines, and subways, residents may feel the vibrations, hear the loud sounds, see the routine traffic patterns, and breathe in the contaminants of the air.

There are a few elements to understand about your site before looking at the details of your specific location. The first one is understanding the air quality of the site. The Environmental Protection Agency has standards for the acceptable levels and types of contaminants allowed. One of the goals for these design solutions is to clean the air to at least the EPA standard, or even cleaner with the use of plant materials. Understanding the quality of air located near airports, expressways, highways, freeways, train lines, and subways means understanding what is in the air and the effects of those contaminants on animals, children, and adults.

 Excerpts from research by Federico Karagulian, et al., 2015, titled "Contributions to Cities Ambient Particulate Matter (PM): A Systematic Review of Local Source Contributions at Global Level" helps to illuminate on air quality in urban areas. The research covers the time period between 1990–2014.

There are five categories of main sources of PM (ambient particulate matter) commonly found that have been used for the purpose of this analysis: traffic, industry, domestic fuel burning, natural sources including soil dust (resuspended) and sea salt, and unspecified sources of pollution of human origin.

- Traffic includes different kinds of emissions from various vehicle types. In addition to primary PM emissions from exhausts, and the emissions of organic and inorganic gaseous PM precursors from the combustion of fuels and lubricants, vehicles emit significant amounts of particles through the wear of brake linings, clutch, and tires (Amato et al., 2009; Belis et al., 2013). These are deposited onto the road and then re-suspended by vehicle traffic together with crustal/mineral dust particles and road wear material.
- Industry is a category including mainly emissions from oil combustion, coal burning in power plants, emissions from different types of industries (petrochemical, metallurgic, ceramic, pharmaceutical, IT hardware, etc.) and from harbor-related activities. Industrial sources are sometimes mixed with unidentified combustion sources or traffic (Belis et al., 2013).
- Domestic fuel burning includes wood, coal, and gas fuel for cooking or heating such as in apartment complexes.
- Natural sources including soil dust and sea salt dust. These components of PM are associated with the resuspension from fields or bare soils by local winds. When reported separately, road dust was included in the traffic source category in this study (Belis et al., 2013).
- The "unspecified sources of human origin" category mainly includes secondary particles formed from unspecified pollution sources of human origin. Primary particle emissions include mechanically generated particles and primary carbonaceous particles. Primary particles also include carbonaceous fly-ash particles produced from high temperature combustion of fossil fuels in coal power plants.

Additional PM are formed from reactions of gaseous pollutants, VOCs (volatile organic compound), NMVOCs (non-methane VOC), and evaporation of solvents such as paints, degreasers, and stain removers.

"Airborne particles like dust, soot and smoke that are less than 2.5 micrometers in diameter are small enough to lodge themselves deep in the lungs. Studies have linked pollution of this sort to respiratory problems, decreased lung function, nonfatal heart attacks and aggravated asthma, according to the United States Environmental Protection Agency." (New York Times Article, "A Study Links Trucks' Exhaust to Bronx Schoolchildren's Asthma," By Manny Fernandez, Oct. 29, 2006).

According to the New York State Department of Health and Department of Environmental Conservation, "particles in the PM 2.5 size range are able to travel deeply into the respiratory tract, reaching the lungs." Exposure to fine particles can cause short-term health effects such as eye, nose, throat and lung irritation, coughing, sneezing, runny nose and shortness of breath. Exposure to fine particles can also affect lung function and worsen medical conditions such as asthma and heart disease. Additionally, EPA officials said these fine particles, a significant portion of which are produced by diesel engine emissions, lead to 15,000 premature deaths a year nationwide.

To understand the scale of 2.5 micrometers, there are 25,000 microns in an inch. The width of PM 2.5 is approximately 30 times smaller than that of a human hair. PM 2.5 is so small that several thousand of them fit on the period at the end of this sentence.

In addition to necessary policy changes addressing gaseous and chemical release into the air, this chapter explores using plants to remediate particulate matter (PM) in the air. Some plants are able to take in the contaminants through their roots, stems, or leaves and continue to live, removing the PM from the air so animals, children, and adults can breathe in cleaner air with less PM.

 The next item to address is sound. Urban sound is measured in dB(A), which stands for decibels with an A-weighting. This A-weighting includes an adjustment which takes into account varying sensitivities of the human ear to different frequencies of sound. A-weighted measurements consider the loudness, annoyance factor, and stress inducing-capability of noises with low frequency and moderate or high volumes. This also takes into account the very low and the very high frequencies and volumes of noise.

Why is this worth understanding? Understanding that sound near airports, expressways, highways, freeways, train lines, and subways have a higher dB(A) allows one to consider the next step of how to use design to lower the dB(A) for those who live or work in these areas.

 Published research by Maarten Hornikx illustrates that plants can reduce the dB(A) ranging from 4 dB(A) to 7.5 dB(A). More information on this research can be found in the reference list. Dense foliage absorbs frequencies, while hard surfaces scatter them. Consider a sound frequency hitting a hard wall, or tree trunk—the sound waves scatter. However, with the softer dense surface of foliage, the frequencies are absorbed. The articles states that:

> Vegetation in urban areas has a range of ecological advantages. Vegetation acts as noise reducing possibility in inner city environments, in particular building envelope greening measures. Also, there is growing evidence that visibility of vegetation by itself affects noise perception positively. For inner city environments, vegetation types that can be considered are low-height noise barriers, vegetated façades, vegetated roofs and trees. Low-height vegetated noise barriers were shown to be useful in road traffic noise applications at street level. These devices can be placed close to the driving lanes, thereby yielding road traffic noise reduction of about 5 dB(A).

Some municipalities, such as New York City, require trees as part of their development in order to assist with noise, views, and air quality. Planting trees in New York City sidewalks has been the project of artists such as Alan Sonfist and Joseph Beuys. Even today, walking through New York City, these living artworks may be unrecognized by the community, yet, without these green artworks, the urban area would have inferior air quality and would lack shade. Other benefits of street trees for a community include experiencing seasonal change, growth, and time changes through plant materials, as well as bringing people together to care for the living plant materials.

In looking at greenwall surfaces, it was found that the greenwall that does not have plant materials on it only reduces 1 dB(A), while a greenwall with plant materials reduces 5 dB(A). It is the soft absorption of the foliage which reduces sound.

Green roofs can decrease sound waves over buildings. It is the combination of plants, the sublayers of soil and liners, as well as roof shape which produce the benefits of sound reduction. It has been found that buildings with vegetated roofs can have an effect of up to 7.5 dB(A) noise reduction. Maarten Hornikx writes:

> The effect of green-wall systems is larger for roadside courtyards than for trafficked streets and may amount up to 4 dB(A). An interesting application of vegetated surfaces are openings to courtyards, greening those surfaces has shown to amount to 4–5 dB(A) reduced sound pressure levels in the courtyard.

What does 5 dB(A) really feel like in volume and effects? Is it really significant? Yes! An increase of 10 dB(A) feels twice as loud to the human ear. For example, with a dishwasher with 50 dB(A) versus a dishwasher with 60 dB(A), the second dishwasher sounds twice as loud at the first one. 10 dB(A) feels twice as loud. Therefore, a decrease in dB(A) using plants and green design by 8 dB(A) feels almost half as loud.

The result—the more green, the lower the dB(A), which reduces hearing loss and stress, while creating healthy environments in an urban area creates positive outcomes.

 One final element to consider is the psychological benefits of views of green spaces and other designs which connect people to green nature. The benefits of green spaces not only provide environmental benefits in negating urban heat, offsetting greenhouse gas emission, and offsetting storm water, there are also health benefits of plant materials. From the University of Washington's *Green Cities: Good Health* (Wolf and Flora, 2010), further mental benefits of green design include:

- The experience of nature helps to restore the mind from the mental fatigue of work or studies, contributing to improved work performance and satisfaction.
- Urban nature, when provided as parks and walkways and incorporated into building design, provides a calming and inspiring environment and encourages learning, inquisitiveness, and alertness.
- Green spaces provide necessary places and opportunities for physical activity. Exercise improves cognitive function, learning, and memory.
- Outdoor activities can help alleviate symptoms of Alzheimer's, dementia, stress, and depression, and improve cognitive function in those recently diagnosed with breast cancer.
- Contact with nature helps children to develop cognitive, emotional, and behavioral connections to their nearby social and biophysical environments. Nature experiences are important for encouraging imagination and creativity, cognitive and intellectual development, and social relationships.
- Symptoms of ADD in children can be reduced through activity in green settings, thus "green time" can act as an effective supplement to traditional medicinal and behavioral treatments.

Further information on the studies mentioned here can be found in the references. From the World Health Organization's, (2016) *Urban Green Spaces and Health—A Review of the Evidence*, their findings summarize that the

available evidence of beneficial effects of urban green spaces, such as improved mental health, reduced cardiovascular morbidity and mortality, obesity and risk of type 2 diabetes, and improved pregnancy outcomes. Mechanisms leading to these health benefits include psychological relaxation and stress alleviation, increased physical activity, reduced exposure to air pollutants, noise and excess heat.

Adding green design for a healthy environment improves the air quality, volume of sound, physical health, mental health, and energy for people who live and work in areas near airports, expressways, highways, freeways, train lines, and subways. In addressing all these layers in a green design solution you can achieve a healthier environment. The users of the site can improve existing conditions of a growing industrial, urban, or changing site by looking at green solutions in both long-term and short-term, as well as in large scale site solutions through room-scale green designs.

WHAT TO DO NEXT

Questions to Ask about Your Site

When looking at sites which are located near airports, expressways, highways, freeways, train lines, and subways, what do you do next? The site solution must offer solutions for both the short and long term as well as addressing small- and large-scale sites.

Studies by Harlem Hospital, Harlem Children's Zone, New York University, and the Institute for Civil Infrastructure at NYU's Wagner Graduate School of Public Service have shown that five of the ten areas with the highest levels of asthma in children in the New York City area are within the Bronx. More information on this article and research can be found in the reference list. The children and adults who attend urban schools are negatively affected by the urban air. Addressing the problem in a variety of ways that you will see in this chapter, including green designs, can improve the health of the site.

Many urban schools are also close to major expressways, industrial areas, subways and train lines. The cluster of urban road, highway, and industrial pollution is in the urban air, the same air which is in urban playgrounds, neighborhoods and schools. Youth in this area have a developing immune system which is affected by the air quality they breathe in routinely. The development and health of the youth are negatively affected by the urban air.

"The study also examined the proximity of expressways to schools. Four expressways—the Cross Bronx, Major Deegan, Bruckner and Sheridan—and the Bronx River Parkway run through or around the South Bronx. About one-fifth of all students from prekindergarten to eighth grade in the area go to schools located within 500 feet, or about two blocks, of major highways, the study showed" (New York Times Article, "A Study Links Trucks' Exhaust to Bronx Schoolchildren's Asthma," Manny Fernandez, Oct. 29, 2006).

There are outdoor air contaminants which get inside through windows-—such as car exhaust fumes, cigarette smoke, and outdoor urban air elements. Urban air and air near freeways, highways, urban roads, airports, and train lines typically contains a higher concentration of contaminants and PM (particulate matter) such as SO_2 (sulfur dioxide), NO_2 (nitrogen dioxide), lead, carbon monoxide, and ground level ozone. These contaminants are a result of vehicle engines, fossil fuels, industrial production, and natural processes such as wind-blown dust. The fine particulate matter of these contaminants can be so small that it is easily inhaled into lungs. This is one of the findings in studies about the high rate of asthma among urban youth.

While laws and policies are long-term solutions for addressing diesel fuels and their use in areas, more immediate solutions are addressed in this chapter. Current air quality issues in urban areas can be addressed by using plants in buildings and neighborhoods located near highways, freeways, urban roads, airports, and trains. Improved air quality helps those who live or work in these areas—people of all ages.

 Addressing the sites which are affected is done on many scales, from smaller scale green rooms, a larger system of green building solutions, an even larger scale of a full block and neighborhood site, and finally through long-term changes in regulations through policies and laws.

One way of considering the scale of the design can be seen below:

Room scale:	Ranges from classroom, gym, office, studio, kitchen, library, dining room, bedroom, lounge, bathroom, and more.
Building scale:	Entry, exit, roof, walls, windows, yard, playground, driveway, sidewalk.
Neighborhood and community scale:	Sidewalks, roadways, road medians, alleys, trash locations, walkways, street trees, gardens, walls, views, freeways, plane routes, public transportation routes and types, amenities, events.

The scale and plant choices of the designs are determined based on the square feet, cubic feet and existing conditions of the site. Analysis includes looking at lights, sun direction, available spaces, electric sources, water sources, and the air quality based on contaminants from the materials presently existing in the site. The scale and aesthetic of the designs are created for each unique site. The designs can reflect local materials and regional plants.

These illustrations of green designs display the integration of a unique, unified application for a community, varying in scale. For example, a green alley area allows people to sit, talk, or use the space to work during their day. Another detail may include a design for planters along a hallway. Additional green designs may include vertical greenwalls, layered plants, and sculptured forms. Using daylight bulbs for a pure white light lessens eye strain. Additionally, there are bulbs and apps which change light to mimic lighting throughout a day from sunrise through sunset. These can help the plants to grow with routine light.

Other elements which can be used include composting. A list of items which are compostable is included in Chapter 3. Depending on the scale of a household, business, or neighborhood, composting can lead to a compost area which is educational and creates healthy nutrients for a site or many sites. The healthy products of compost can help replace nutrients in the soil as needed throughout the lifespan of a site. For example, compost created from a neighborhood can be used in a community garden, for the community to use in their personal sites, or to sell to help raise funds for the community.

The creation of furniture and accessories which contain living plants, such as those illustrated in Chapter 1, can not only improve the air quality and mentally improve the users, but also create jobs for people. A neighborhood group, students, artists, community members, or people who are learning a vocation can create a series of designs for use at their own site, or to sell to benefit the community.

Re-using stormwater on a neighborhood scale, urban building scale and house scale not only has tax benefits, but also reduces costs for water usage and connects a group of people with a shared purpose of creating a healthier site. Communities can be stronger by sharing green design practices and working together for a healthier site.

 This closing chapter—longer than the others—looks at the room scale, building scale, and neighborhood scale for two different sites near highways, urban roads, trains (below and above ground lines), and airports. The two examples in this chapter include a charter school in the South Bronx near multiple freeways, and a rural house in Minnesota adjacent to a freeway. There are many other sites in which the principals of these sites also apply. As in the previous chapters, Ashley Kaisershot (AK) continues her insightful questions to illuminate topics for clarity, understanding, and inspiration.

Example A—Charter School for International Cultures and the Arts Located in the Bronx, NY

An urban neighborhood

- This diverse neighborhood is located at the intersection of two major freeways — interstate 87 and interstate 278.
- There are three bridges which connect this neighborhood, the Port Morris neighborhood, to other areas of New York City.
- There are over 67,000 people living in the Port Morris neighborhood of the South Bronx (www. statisticalatlas.com), with more people working in the area throughout the day and night.
- In the neighborhood, there is a school, playground, street parking, and industrial buildings currently in use by businesses.
- Small businesses are located in mixed-use buildings with small shops on the bottom floors and residential housing above the ground floor.
- High-rise residential apartments are also located throughout the neighborhood.
- The air quality is directly affected by the freeways and industrial businesses located in the area.
- The conditions of urban sounds and urban air quality exist.
- Urban air typically contains a higher concentration of contaminants and PM (particulate matter) such as SO_2 (sulfur dioxide), NO_2 (nitrogen dioxide), lead, carbon monoxide, and ground level ozone.
- These contaminants are a result of vehicle engines, fossil fuels, industrial production, and natural processes such as wind-blown dust.

Images 5.1, 5.2, 5.3.

South Bronx, New York, five-story school building on a major street located at the intersection of Major Deegan Interstate 87, Interstate 278, and near three bridges.

The diverse neighborhood has a school, industrial buildings, small businesses, a playground, park, and apartment complexes.

Image 5.1 by Google

An urban school building...

- The five-story building opened as a charter school in 2014.
- The playground across the street from the school is used by neighborhood and school children.

Image 5.4.

Charter school in the South Bronx, New York. The building is 40,000 square feet, with five stories. The neighborhood has businesses which use warehouses for manufacturing and storage. There is also an area for the NYC Department of Sanitation.

- The playground has a fenced court used for handball and other activities.
- The playground is surrounded by the school, warehouse businesses, a Department of Sanitation location, underpass, a bridge, and multiple interstates and freeways.
- Located in a triangle of land where multiple streets meet, the 2014 school building is LEED certified.
- There are approximately 400–450 students who attend the school from kindergarten through to fifth grade.
- There are school buses which pick up children throughout the area. When sitting idle, the school bus exhaust can directly affect the air and the people standing on the block waiting with the children and staff.
- The majority of the windows in the building do not open. There are only a few windows which are designed to open.
- In addition to urban scenes, there are deciduous street trees planted around the building sidewalks.
- There is natural light through the day in the rooms.

Images 5.5, 5.6, 5.7.

Bronx, New York.

The library, classrooms, cafeteria, performance room, and hallways have large windows with views to the neighborhood.

Classrooms, offices, and common rooms...

- There is a large room which serves as both the cafeteria, school meeting room, and performance space with a stage.
- The library is designed for younger children with lower height shelving, chairs, and tables.
- The hallways are approximately eight feet wide and have student artwork throughout.
- The classrooms are on both sides of the hallway and have windows facing the urban scene.
- The ratio of students to teachers is approximately 20:1.
- Classrooms have tables for multiple students to work at, rather than individual desks. Dance rooms, technology rooms, and art rooms have an open floor plan with furniture as needed.

The Green Design Can Improve

Knowing all these factors, the resulting design can improve the air quality of the neighborhood with plants which can help clean the air of <u>urban air</u> contaminants. The playground design can also have planted fences to help with <u>sound and air pollution</u> from the neighboring businesses, roads, interstates, and bridges. Green roofs on the surrounding <u>industrial buildings</u> create a pattern of plants in the neighborhood to improve the short and long term of urban life. The improvements include both the <u>psychological effects</u> of viewing plant materials as well as <u>environmental effects</u> of plants which clean air contaminants. Using the <u>grey water</u> and <u>storm water</u> from the school site for additional outdoor plants, as well as grey water for interior designs will help with long term sustainable success. Green columns at the entrance help improve the <u>air quality affected by the car and bus exhaust fumes. Electric composters</u> inside the school can be utilized for <u>food waste</u>, and natural cleaners can replace <u>complex chemical cleaners</u>. The hallways, cafeteria, assembly area, classrooms, and office use <u>paints, cleaners,</u> and <u>sealants</u> which affect the air quality. The air quality can be improved throughout the site with greenwalls and structures.

Example B—Two-Story House in a Rural Neighborhood
Adjacent to the Freeway in Little Falls, MN

A neighborhood near the freeway

- The neighborhood consists of approximately twelve residences, in a low density residential setting within the rural community.
- The lot sizes vary in acres from half an acre to two acres.
- Nearby roads include a gravel street.
- The property is separated from the highway by a gravel road and right-of-way fence.
- There is a steep slope from the freeway down to the houses.
- It is community of people who routinely commute to their work site.
- The emissions from the freeway include those from vehicles which use both ethanol and diesel fuels. Air near major roadways typically contains a

Image 5.8.

Little Falls, MN.

The four-lane freeway runs adjacent to the neighborhood.

Image 5.8 by Google

higher concentration of contaminants and PM (particulate matter) such as SO_2 (sulfur dioxide), NO_2 (nitrogen dioxide), lead, carbon monoxide, and ground level ozone.

Images 5.9, 5.10.

Trees are planted to block the view of the freeway from the house. The owners are currently a young couple who are renovating it.

Images by A. Kaisershot

A rural property…

- It is a traditional Midwest single family rural home with a large front yard, side yard, and minimal rear yard.
- The plant materials on the property are many white pines, elms, and spruces used to separate the view of the gravel road and fence adjacent to the right-of-way of the freeway.
- Turf grass is the primary ground cover.
- The building footprint is about 1,200 square feet on a half acre plot.
- The house is two stories built in 1920 with concrete, wood, plaster, glass, metal, and shingles.
- The house was relocated and moved onto the property in 1992.
- A new foundation was created for the house on the property.
- The two-story building is approximately 30 feet ×50 feet with a roof slope of 9:12.
- One source of the contaminants of this house includes emissions from hardwood floors which were replaced and resealed.

The Green Design Can Improve

Knowing the details of the site, the resulting design can improve air quality in the neighborhood by using plant materials which remediate the contaminants from car and truck exhausts. The neighborhood can have common areas along sidewalks and greenways which aesthetically connect the community as well as functionally improving the air quality from the adjacent freeway which affects the community. Additional design solutions for the house include being cognizant during the renovation of what chemicals are in the wood strippers, paints, cleaners, and sealants being used. Air circulation during and after renovation can also be used for a long-term healthier site to improve the contaminants in the daily household cleaners, furniture, and cabinets. Composting and recycling capabilities can also be added to the site for a long term healthy lifestyle choice.

Images 5.11, 5.12, 5.13, 5.14, 5.15.

Little Falls, MN.

The yard wraps around the house. The house is in the process of being renovated and restored to its original character with modern amenities.

Images by A. Kaisershot

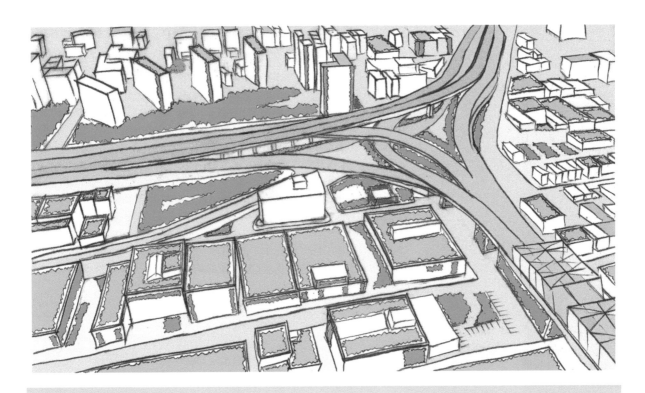

Images 5.16, 5.17, 5.18.

Community design for an urban area with a public school, warehouses, residences, freeways, bridges, and businesses. This design incorporates adding plant materials throughout public and private spaces to create a connected healthy urban experience. The addition of planted walls, green fences, street plants, planted alleys, green roofs, and other unique ideas enhances the thriving urban experience with new approaches to park design. The new park ideas include roof parks across multiple buildings, alleys for green parks, and including green vertical designs throughout the parks.

Image 5.17 by Google.

GREENING THE URBAN NEIGHBORHOOD

Images 5.19, 5.20.

Creating a plan which adds 20% more plant materials to a site is the goal of this design. Additionally, this design must address choosing plants which will grow in an urban area, as well as remediate urban air contaminants.

In this design, the urban area includes billboards, sidewalks, alleys, buildings, fences, and other elements with trees, shrubs, grasses, and vines.

Chapters 1 and 2 list plants for urban areas and car emissions, as well plants which are also listed in this chapter. The database for plants which are used for remediation can be found at www.engaginggreen.com.

The diversity of the neighborhood is an asset and can include designs which are vertical, horizontal, hydroponic, soil based, have original aesthetic form, and a healthy function.

bridge and interstate with planting on walls

planted wire fence around playground

greenwalls in underpass watered by stormwater

permeable pavers on walkway

greenwalls on underpass & bridge area along pedestrian walkway

PLANTS FOR AIR QUALITY

Green design can be used to connect the neighborhood while also encouraging individuality.

Plants which remediate car emissions include,

Buffalo Grass (*Bouteloua dactyloide*)
Blue Gamma Grass (*Bouteloua gracilis*)
Bamboo Palm (*Chamaedorea seifrizii*)
Orchid (*Dendrobium taurinum*)
Red-Edged Dracaena (*Dracaena marginata*)
Rubber Plant (*Ficus elastica*)
Weeping Fig (*Ficus benjamina*)
Gerbera Daisy (*Gerbera jamesonii*)
Boston Fern (*Nephrolepis exaltata*)
Kimberly Queen (*Nephrolepis obliterata*)
Heart-Leaf Philodendron *Philodendron scandens*)
Dwarf Date Palm (*Phoenix roebelenii*)

AK: *Can I create a* **mobile design** *to block views for privacy?*

SF: Yes, this can be quite easy to do. In Chapter 3 there is one example of using aeroponic plants to create a sliding lace curtain over the window that can obscure the view. The more plants, the more one can obscure or block the view. Making it mobile, in this case as a slider on a curtain rod, allows you to change it in the changing light or to clean the window, or move it if needed for any other reason.

Mobile designs can also be on wheels or sliding elements. This can include planters, walls, sculptures, curtains, artworks, and anything the mind can think of.

GREENWALLS ON THE BUILDING EXTERIOR

Image 5.21.

The building design has greenwalls which are watered from storm water collected on the roof and piped down the walls. Grey water from the sinks throughout the building can also be piped to the greenwalls, as well as water dripping from air conditioners.

GREENWALLS AT THE BUILDING ENTRANCE

Images 5.22, 5.23.

Greenwalls on the building utilize grasses and vines which remediate air contaminants near the freeway. In addition to greenwalls on the exterior of the building itself, green columns at the school's main entrance are decorative and functional.

With the goal of adding 20% green design per square foot, it is important to add plant materials where the exterior air mixes with interior air. It is in those spaces where the plants can be even more effective in assisting the air cleaning system to create a healthier site.

AK: Can you explain PM (**_fine particulate matter_**) in more detail?

SF: Particulate matter, PM, is how the size of air contaminants is measured. Let's look at it this way—imagine a boulder, a rock, a pebble, and a grain of sand. These four scales range from boulders larger than you can wrap your arms around; rocks you may need both hands and part of your arms to hold; pebbles that you can fit a quantity of in your hand; and sand which you can hold in uncountable amounts in your hand and the grains are fine enough to slip through your fingers.

PM is similar in that is has scales which define if they are large enough to see with the naked eye or small enough that you can breathe them into your lungs. Fine particulate matter is of the most concern in air quality because that is the size that gets into our lungs. For people that have been near a fire or smoke, they know the feeling of breathing some smoke into our lungs. Smoke contains PM which is unhealthy for us to breathe. That is why we naturally cough in an effort to get it out of our lungs. There are some PM which are also unhealthy and we breathe. When an unhealthy amount of it is in our body it can cause damage.

*AK: Does **air in rural** areas also have pollutants?*

SF: Yes, air in most areas has pollutants. The air contaminants vary based on the site conditions. This may include sites such as agricultural food farms with pollutants from pesticides as well as gasses from decaying byproducts of animal farms; industrial pollutants from manufacturing plants; and other contaminants based on the site conditions.

There are studies which indicate the contaminants which are typical for different types of sites, such as urban, industrial, agricultural, rural, suburban, and other sites. The air at a site can also be tested. The contaminants may vary based on changes in the area in construction of various buildings, wind patterns, and other common elements.

OUTDOOR PLAYGROUND

Images 5.24, 5.25.

Those people in the outdoor playground experience the effects of loud street sounds and urban air. The first step is to replace fences with greenwalls to grow plant materials which can be irrigated by rain as well as building grey water. The greenwalls create a visual barrier as well as reducing the noise levels in the playground.

Additional designs include a permeable ground cover with grasses growing in between hardscape stones. Where the playground meets the building edge, the design of pipes is used to pipe the water from the collected water on the roof to water the playground plant materials.

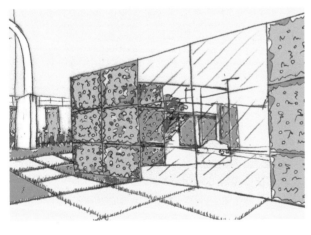

*AK: What role do plants play in **phytoremediation**?*

SF: Phytoremediation is the use of plants to remove contaminants from the soil, water, or air, and falls under the general term of remediation as a specific type.

Though there are multiple ways plants can be effective in phytoremediation, generally, the most common way is for the plant to uptake the contaminants through the roots of the plants. Other ways include taking the contaminants in through the stem or leaf of the plant. When plants are able to uptake the contaminant through the root it is more effective and beneficial because the plant can then filter the contaminant before it rises up to the stem or leaf of the plant.

For example, if the contaminant of lead rises to the stem and leaf of the plant, the plant still lives well. However, when the leaf naturally drops off and decays back into the soil, the lead that was in the leaf—no matter how minute—goes directly back into the soil. When the lead is held in the living, growing roots, the lead will stay there for as long as the plant is growing and living.

Plants can uptake the contaminants into the roots whether they are planted in soil, growing in water, or are air plants.

HALLWAY DESIGN

Image 5.26.

Hallways are a space of continual movement with large crowds of students and faculty in a common shared space. The movement in that space during the change of classes continually stirs up dust and air contaminants. This green design uses plants to clean the air of contaminants as they are moved around while groups of students are walking between classes.

AK: *How do plants work to __improve the air quality__, specifically for those that uptake particular contaminants?*

SF: That is a fantastic question. All plants clean the air. No matter the size of the plant, their love of direct light or low light, whether they are growing indoors or outdoors, their ability to flower or not, or whether they have foliage year-round, they can all help clean the air.

There are some plants that can take in specific contaminants because of the chemical make-up of the plant. For example, research has shown that sunflowers can take in lead and live well, continue to grow, bloom, and return yearly with lead in the roots of the plant. There are other plants that do not live or thrive with the presence of lead. Those plants may either reject the lead and refuse to take it into its system, or generally not grow as well due to the presence of lead.

This is the same for other contaminants and plants. There are plants mentioned in this chapter which can take in biodiesel fuels and oils, and live well. Some plants even chemically break apart the contaminant to more basic elements. Research continues in labs and in natural conditions to see which plants can take in contaminants. Contaminants vary from organic, biodiesels, oils, metals, radioactive gases, and others.

AK: At what **distance** from the source of the contaminant is my building or neighborhood **affected by air quality**?

SF: There are several factors in determining the distance a contaminant can spread from the source. However, the short answer is that is can be for quite a distance, even miles. Some of the factors which are used to determine the distance include the wind current, the wind pattern, the structures which may be blocking the air movement (such as walls or buildings), the slope of the land, the current weather, and the contaminant itself.

Let me help to break this down a little bit further. Some contaminants are larger or heavier and drop to the ground surface quickly without moving with the wind. In areas of rain or heavy humidity, the contaminant in the air may get attached to the humidity in the air and be weighed down, forcing it to fall to the nearest ground surface. Once on the ground surface, such as a road, the contaminant may stay where it is at or move down to the lowest point of the slope. It may move to the lowest point of a slope, no matter how gentle the slope, with the help of running water, street cleaners, people walking and carrying it on their shoes, or natural gravity.

Wind also carries air and all that is in it. Whether it is forceful or gentle wind, air is moved short and long distances. There are items which create wind, such as a moving truck. If driving in a small car and a large truck has driven by you, you have likely felt the force of the moving truck. It is not just weather that creates wind, but also moving forces. Some of you have experienced seeing the fur of your pet floating in the air, then sitting on the wood floor, then moving in the air again when you walk by it, causing enough of a movement from your foot nearby it for it to again move into the air and land a few feet away. Some contaminants in the air can be understood in the same way.

Wind patterns change based on structures. When I lived in Fargo, North Dakota, one alley routinely had a noticeable wind tunnel and even had a large gathering of snow during the winter season. The snow drift was an indicator that the buildings were altering the wind pattern such that the force of the wind was stronger in that area of the alley.

CLASSROOM DESIGN

Image 5.27.

The administration and faculty of school systems can benefit both physically and emotionally from plants in their space. Chapter 4 addresses the psychological benefits of greening. In this site, teachers and administrators are on-site daily. Green designs can inspire those adults who are in charge of inspiring children.

The greenwalls here can be mobile columns which are moved as needed for changing room arrangements. Plants on these columns are both for air quality as mentioned throughout this chapter, as well as ones with colorful blooms to show temporary changes throughout the school year.

Essentially, wind moves naturally, hits an unmovable source such as row of buildings along a block, the wind finds a gap of an alley between buildings and the wind goes in it strongly, then is caught in the alley with the snow it was carrying. The wind carries the snow to the alley along the side of the building, and deposits it there. Some of the wind escapes the alley. You can feel this when you walk across the front of the alley and feel a large gust of wind that is not as strong in other areas along the same block.

The same applies for the air when there is no snow, but rather there are contaminants in the air from car exhausts, paint fumes, construction debris, etc.

CREATING PLANTED DESIGNS WITH THE STUDENTS

Image 5.28.

As the plants continue to grow, students can work with the plant growth to create new living designs. The new designs can be kept on-site or brought home to continue the healthy lifestyle in their home.

Images 5.29, 5.30.

This design adds plants to the community to improve the air quality while improving community relationships and enhancing the aesthetic design. Little Falls, MN.

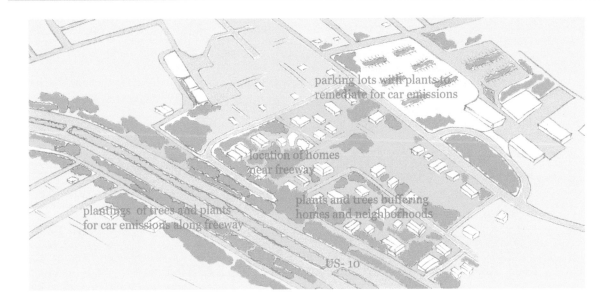

MAP OF THE NEIGHBORHOOD WITH GREEN DESIGN INTERVENTIONS

A master plan of the neighborhood illustrates the many common spaces and private spaces which can be used for planting. The plants chosen for these spaces are those which can remediate for the contaminants of car emissions.

NEIGHBORHOOD PLANTINGS

Image 5.31.

Plants which remediate car emission contaminants include:

Spider Plant (*Chlorophytum comosum*)
Chrysanthemum (*Chrysanthemum x morifolium*)
Warneckei (*Dracaena deremensis 'Warneckei'*)
Red-Edged Dracaena (*Dracaena marginata*)
English Ivy (*Hedera helix*)
Lady Palm (*Phapis excels*)
Dwarf Date Palm (*Phoenix roebelenii*)
Snake Plant (*Sansevieria trifasciata*)
Peace Lily (*Spathiphyllum wallisii*)

AK: *Can you explain how* ***plants*** *help* ***absorb sound*** *in more detail?*

SF: Plants can absorb sound through their leaves and foliage. Dense trees are more effective than trees in large spaces without greenery. The bark of trees is a hard surface which sound waves bounce off and scatter. The green areas of plants are soft and have pockets of air. When sound waves are directed to the foliage, the sound waves are softened and absorbed into the leaves and air pockets, similar to a soft sponge. This is effective for urban areas where there are many sounds which move through

Neighborhood-scale designs include incorporating street trees, shrubs, grasses, and planted fences. Rain gardens are also incorporated throughout the neighborhood to utilize storm water as an amenity in the community.

This example specifically includes designs which are used to address the proximity of this house to the freeway. In Chapter 3, elements of composting, recycling, privacy and more are reviewed. Many ideas from that chapter can be applied to this site, in addition to designs which are for the specific challenge of this neighborhood from the adjacent freeway.

OUTDOOR SPACES, BUFFERS, AND SCREENS

In this design, plants and hardscape are used to screen the amount of sound and light that can carry from one property to another. Buffers can also ensure attractive views from a property and to neighbors. Appropriate landscaping also moderates temperatures of impervious areas, reduces glare from parking lots, and helps filter automotive exhaust fumes.

Buffers and screens are used to restrict the view of adjacent areas, which can be significantly different character, density, or intensity. A buffer consists of a horizontal distance from a lot line. This buffer can consist of screening, landscaping, and fencing materials. Screens can be made from landscaping, fencing, berms, or combinations.

<u>HOME WITH A PITCHED GREEN ROOF</u>

Image 5.32.

Putting plants on a roof, whether a sloped roof or flat, keeps the roof temperature from fluctuating too severely in the extreme weather conditions of a hot summer and cold winter. The planters keep the roof and home more insulated, keep the temperatures from extremes, lowering the cost of heating and cooling in long term.

the air. When dense areas of plant materials are in front of buildings, properties, courtyards, yards, entrances, or windows, the plants act as sponges for sound waves. These plantings reduce the sound that travels to residents in the building, property, courtyard, yard, or room.

AK: Are **noise reduction windows** *effective for soundproofing, even in urban areas?*

SF: Windows which are designed for noise reduction can improve the ability to have a more quiet building. Owners may love the view and lifestyle of an urban life, yet do not want the noise pollution of cars, trucks, music, and other urban sounds. There are multiple companies which produce noise reduction windows, some with an effectiveness of 80–95% noise reduction. There are a few elements to understand when exploring this option.

OUTDOOR SPACES

The exterior plant design uses form and function to create a welcoming entry in all seasons. While the form of the design is aesthetically pleasing, the function improves the health of the site.

The air, soil and storm water in the site has contaminants which come directly from the vehicles on the nearby freeway.

Similar designs can be used in all properties in the neighborhood to benefit the health of the community.

Noise reduction windows are rated with STC scale sound transmission class. The higher the number, the more effective the window is at barring sound. A typical single pane glass window has an average STC rating of 27, while dual pane windows have a 28 STC rating. Soundproof windows are rated at least 45 STC and can be as high as the mid-50s. STC ratings in the mid-50s can block as much as 95% of noise.

Noise reduction windows use thicker glass, increase the distance between the window panes, and use laminated glass which is a glass/plastic/glass layered design. This design reduces noise transmission.

There are other options in blocking sound, such as fabric which absorbs sound, caulking around doors and leaks, and insulation. Another thing to remember is that there are different sound frequencies. Low frequency sound such as the rumble of a large truck, are more difficult to block than the high frequency of birds chirping.

ENTRY DOOR OF THE HOME

Image 5.33.

The entry not only creates the mood for a space, it is also typically where the most air from the outside mixes with the indoor air. This can occur at the front, side, or rear door of a home.

This entry has green-walls at the doors to the home for quickly addressing the air which comes through when the door is open. A similar concept can be created for areas near windows which are typically opened.

The sound may also be coming through air vents, chimneys, leaks in windows or other sources. For example, brick is more effective at sound reduction than wood. There are professional acoustical consultants who can work with you on your site and needs for the most effective solution.

AK: At what **distance** *from the source of the sound is my building or neighborhood* **affected by sound volumes***?*

SF: Similar to the answer about air contaminants, there are several factors in determining the distance of the source of the sound to see how far it can spread. The same factors include the wind current, the wind pattern, the structures which may be blocking the air movement (such as walls or buildings), the slope of the land, the current weather, and the frequency and soundwaves from the source itself.

The factors of wind, wind pattern, slope, and weather are similar to the earlier question. Some additional elements which affect sound include the materials of the structures and frequency of the sound, as well as its volume.

WINDOW DESIGNS

Image 5.34.

In Chapter 3, curtains made of air plants are shown as a green design.

The addition of planters along the floor with tall grasses blur the materials of interior and exterior areas. This allows the residents to enjoy the scent and color of grasses in their home, with views to the exterior grasses. The grasses also remediate the air contaminants from the nearby freeway.

Surfaces which are soft, dense, and have air pockets absorb sound. Foliage and sponges can be used as soundproofing material in buildings. The frequency of the sound itself also affects how far it is carried. Low frequency sounds such as bass sounds can shake and vibrate much further than higher frequency sounds such as birds chirping. One other factor in how much the sound affects you is the context in which you hear the noise. For example, trucks moving in the middle of a day along with other rustling traffic is less likely to bother you than a single loud truck of the same frequency in the middle of a quiet night. Though the frequency of the sound is the same, the context shift affects you more.

DESIGN STUDIO

Image 5.35.

The studio for this designer needs wall space so she can hang her designs while in process. With little wall space, this design utilizes the ceiling for the greenery and blooms. Using a flattened planter and plants along one wall, the plant growth along the pitched ceiling remediates the contaminants from the markers and glues, while providing inspiration and health benefits.

SCIENCE BARGE BY GROUNDWORK HUDSON VALLEY IN YONKERS, NEW YORK

Images 5.36, 5.37.

In 2008, this prototype barge was fitted with a hydroponic green-house with a sustainable urban farm which is powered by solar panels, wind turbines, and biofuels, and irrigated by rainwater and purified water.

Images by Donna Davis/Ms. Davis Photography

URBAN FOREST

Image 5.38.

Stephen Glassman's group proposed using large-scale structures for urban forests. New technology can be used to turn existing structures, such as billboards into a connected urban forest. Funded by a Kickstarter campaign, the first billboards are proposed to be created for a Los Angeles, California, freeway.

Image from Open Air.

Stephen Glassman-Artist.

David Hertz-Architect

HEART-SHAPED VINE

Image 5.39.

In Bellingham, Washington, there is a building with a heart-shaped vine. Well-loved living pieces such as this add fresh scent and living art in urban areas.

Image by Max Lowe/Getty Images

*AK: Generally, what views are considered **aesthetically pleasing**? Or what do you look for when considering aesthetic appeal?*

SF: I remember when I was in graduate school studying Landscape Architecture, asking a similar question to a professor. Defiantly, I stated that urban views may be unappealing to some, and pure heaven to others. How can one measure this when there are far too many personalities and preferences? Professor Curry was able to be a clear and effective teacher when explaining what studies have shown to be generally aesthetically appealing. From his answer and research since, I have found the following answer generally works in design.

Research has found that people enjoy pleasing foregrounds, middle ground, and backgrounds. This means what is close in front of us, what is in the middle ground, and what is in the distance. The immediate items in view are most personal and what we feel we are able to affect. These are the spaces we can walk to within a few feet. These can include the items directly in front of us in a room or outside the window, such as a yard, fire escape, or building next to you. These items create a personal space for us and people feel connected to this space. Being able to design the immediate space around us is appealing.

HOTEL IN PICKERING, SINGAPORE

Image 5.40.

PARKROYAL hotel in Singapore is designed by studio WOHA. The vertical gardens are planted on the balconies of the building. Over 15,000 square meters of vegetation are incorporated into this building.

Image by iStock

The next layer is the middle ground. This layer is a space you may be able to walk to or get to by some mode of transportation. For those who live or work on sites less than a few acres, this is typically a public space or at least not your space to design. This space can be the most challenging for people because the sound and sight of it affects our opinion of the space, yet we usually have little control over it.

The final space we look at is the distance, the vista, the long distance view. This may be a skyline, a treeline, flat land, parks, golf courses, the ocean, or any variety of items. This helps to create the atmosphere of a space. This longer view is part of the spaces we personally find appealing as it connects the full picture of our space.

In design we look at which of these layers are appealing, what elements in each layer the client finds appealing, and how to enhance or block elements of layers accordingly. In this book, the intentional

PASONA URBAN FARM, TOKYO, JAPAN

Image 5.41.

Located in downtown Tokyo, Pasona HQ is a nine-story high, 215,000 square foot corporate office building for a Japanese recruitment company. The double-skin green facade features seasonal flowers and orange trees planted within the 3 foot deep balconies. Partially relying on natural exterior climate, these plants create a living greenwall and a dynamic identity to the public.

Design by Konodesigns, LLC.

Image by Toshimichi Sakaki

PASONA URBAN FARM, TOKYO, JAPAN

Image 5.42.

Built in 2010, this 10,764 square foot plant factory is located inside a former major bank vault. This hydroponic farm has a rice paddy and other plants throughout the urban site.

Design by Konodesigns, LLC.

Image by Luca Vignelli

blocking of views, when possible, is done with plant materials. The choice of plant materials is for a larger greener healthier lifestyle.

AK: *Do you look at* **_aesthetics and design_** *differently for sites* **_near airports, expressways, highways, freeways, train lines, and subways_**?

SF: Based on the previous question—not really. In looking through the eyes of a designer, I still look at these sites based on the foreground, middle ground, and background. I also look at the ground plane—the surfaces one walks one; the eye plane—typically created for humans of an average height of 5'7" and what is seen within the eye plane; and the overhead plane—such as trees, arbors, structures, buildings, and sky.

THE LOWLINE LAB, NYC

Images 5.43, 5.44.

The Lowline in NYC is applied research of green technology. Using technology to bring enough light underground to create photosynthesis, the team is working to transform an underground, abandoned subway terminal in the Lower East Side of New York into an underground park.

The Lowline Lab was created to research and test some of the technology and was open to the public from 2015–2017.

THE LOWLINE LAB

The plants in the Lowline Lab are all living, and some are edible. Creating a sustainable ecosystem in the lab has helped the project to move towards a projected 2021 opening of the underground park in the NYC subway terminal.

THE LOWLINE LAB

Images 5.45, 5.46

The diagram illustrates the sunlight piped in through plumbing tubes from the rooftop, using mirrors with lenses to distribute and reflect the light throughout the space onto the ceiling of the Lowline Lab.

People who choose to live or work near airports, expressways, highways, freeways, train lines and subways are aware of it when choosing the space. People may intentionally choose the space because of the proximity to public transportation, quick and easy access to roads, or frequent travel via airplane. Choosing these sites for their appeal does not negate the ability of people to improve the health of the space for those people who choose to live or work there.

AK: At what __distance__ from the unappealing scene is my building or neighborhood __affected by views__?

SF: All distances and planes affect a view. Some things cannot be changed. You may choose to design the space to enhance the beauty of the views and block unappealing views. You may choose to move if the scene does not inspire the best in you. You may look within yourself to find what you find most pleasing and seek that out. Or you may look inside yourself and realize that you can be content by changing your attitude, approach to the space, or a few details such as a fresh scent, to make your feel inspired.

**THE GREEN LINE,
GARDEN PARTY SERIES
(FROM LEFT TO RIGHT)
HERB GOWN,
AIR PLANT GOWN,
LAWN COAT,
OPERA GOWN, AND
WEDDING GOWN**

Image 5.47.

In 2012, the Green Line was created as a wearable landscape, pushing the lines of art, fashion, and landscape to combine them in one series. The plants were living and growing hydroponically through the fabric. The pieces grew over a seven-month period, with daily watering and nutrients.

The Green Line, Garden Party Series. Artist: Stevie Famulari

Image 5.47 by Yvonne Denault/ www.yvonnedenault.com

AK: *What happens if the building has **itstoric meaning or has landmark status**? Does this affect what green practices we can do?*

SF: Yes and no. First, itstory is a word proposed in recent times to be a collaborative narrative in the response of people choosing herstory rather than history (his-story). Itstory combines not only both genders, but also includes animals and plants. It creates a more collaborative narrative and is more inclusive of greater diversity.

A building with a rich itstoric past can still have green improvements. Based on the location of the building and its legal status, one can usually create a variety of designs on the interior without changing the exterior. The exterior is typically what may need to be preserved if the building has landmark status.

You may also be able to put a layer in front of the building of plant materials, greenwalls, or vertical gardens. If these layers do not touch the structure, they are sometimes allowed by the policy of the state. Also, creating plant materials around the property should not affect landmark status. Therefore, plant materials added to the site which are not on a structure are typically allowed. Fences and other structures may also be allowed to be removed, altered, or covered. Check with the policies of your location.

AK: *How do you **phase in green practices** when some are based on a community scale, while others are on a small private scale?*

SF: That is an excellent closing question. Few designs are built on existing sites with all elements being completed at once. Due to budgets, use, timelines of the project, and many other factors, most designs of this scale are phased in.

THE GREEN LINE, GARDEN PARTY SERIES OPERA GOWN AND WEDDING GOWN,

Images 5.48, 5.49.

The plants grown on the gowns include a grass mix, cosmos, marigolds, chives, basil, parsley, mint, cilantro, and air plants. The only plants not grown on the gowns are the red flowers on the wedding gown (left) which were added for the photography shoot. The flowers on the opera gown (above) were grown from seed directly into the gown.

The Green Line, Garden Party Series. Artist: Stevie Famulari

Images 5.48, 5.49 by Yvonne Denault/ www.yvonnedenault.com

A professional team of designers, engineers, architects, planners, scientists, researchers and others create a master plan which has the full scope of the project. Then, based on the client's wants and needs, budget, timeline, usage of the site, traffic and other factors, the team puts together a timeline for the project. The first phase may consist of getting the infrastructure for the full scope of the project in place. The infrastructure may not be seen, but allows the full scope of the project to be completed with all necessary electric, water, and structural needs when it is ready.

The first phase also may involve one or more elements which are eye-catching and pleasing for the users. These brilliant green elements may be used to raise funds for the additional phases or just to have some of the designs of the full project scope completed.

Further stages are then completed on a timeline determined by the team, clients, and contractors. Projects may take a few years to fully complete all phases of a community site.

A FEW NOTES AS YOU ENJOY YOUR GREENING PROCESS

Creating a set of goals may help you to go through the project with more focus and understanding of the direction of your project. Your goals may change as the site changes. However, it is a starting point that helps some people to move forward and be able to work with others towards a successful green design.

The sample set of goals below was created for an interior green project and is still being used today for the long-term success of it.

Short term goals, 2–5 years

- Create airflow and air circulation throughout a room and the entirety of the space.
- Choose plantings for indoor air purification.
- Create greenwalls for cleaning air contaminants.
- Reduce dust.
- Eliminate mold.
- Set up both fresh and grey water systems.
- Set up a circulation design and planted design to work for year-round comfort.
- Renovate a conduction stove top.
- Create a recycling program for the building.
- Create a composting program for the building.
- Switch all lights to daylight bulbs.

Long term goals, 5–25 years

- Maintain air purification, including scheduled testing of the site.
- Remove or seal lead pipes and lead paint.
- Replace gas stove top to a conduction stove top.
- Create consistent and comfortable humidity levels.

- Create consistent and comfortable temperature levels.
- Replace old windows with soundproof windows.
- Renovate walls to create a planted edge for apartment cleaners.
- Set all plants on timers as appropriate for the temperature and humidity.
- Create community space for all people who live in the building to socialize.

This chapter explores green design solutions on a variety of scales—the room scale, building scale, and community scale. This involves a team of people on many sides of the project. All the people of the community or neighborhood can work together to create a design which improves the quality of life for those who live in the area. These clients have a great range of age, cultures, and diverse perspectives. As clients, a group such as this can be the most challenging and the most rewarding. The challenge of working for such diverse users means taking into account their vast range of interests.

However, this can also be the most rewarding client, because the design is lived in constantly by many people. The green design can improve the lives of many people beyond your own lifetime and in ways that were unforeseen when the project was started. The users also include the animals and plants of the site—both of which are living creatures. With a full scope of users, the fuller ecosystem can take into design change, growth, and diverse users.

It is not only the diverse clients that make projects such as these challenging and rewarding; it is also the diverse team of professionals. These diverse teams have professionals who are all knowledgeable about their field, and yet may need improvements to help apply their knowledge collaboratively with the other professionals. Finding the common goals and common ground is a rewarding experience that improves the team for the better of the community.

Having fun as part of the process, feeling proud of the process and getting people involved always helps. Understanding that every voice counts, every user is important, and some things need to shift for the better of the overall design will help everyone to work together toward fulfillment and long-term success.

Image 5.50.

Street in New York City with approximately 10–15% more plant material added per square foot of surface area. There are unique approaches to adding green designs to any space, no matter the scale. The addition of even 10% more plants improves the air quality for a healthier environment for all users of the spaces.

References

Arsenault, P., Darington, A. (2012). Indoor air biofilters deliver clean air naturally. Biological systems function to improve air quality while providing beautiful form. Architectural Record.

Belis, C.A., Pesoni, E., Thunis, P. (2013). Source apportionment of air pollution in the Danube region. Retrieved from http://iet.jrc.ec.europa.eu.

Brumfield, C.R., Goldney, J., Gunning, S. (2008). *Whiff. The Revolution of Scent Communication in the Information Age*. New York. Quimby Press.

Cooper Marcus, C., Barnes, M. (1999). *Healing Gardens*. New York. Wiley & Sons.

Ehrlich, S.D. (2011). *Aromatherapy*. University of Maryland Medical Center. Retrieved from www.umm.edu.

Environmental Protection Agency. (2016). Air quality planning and standards. Retrieved from www.epa.gov.

Fernandez, M. (2006, October 29). A study links trucks' exhaust to Bronx schoolchildren's asthma. *The New York Times*, Region Section.

Georg, M.E., et al. (2011). Thirdhand tobacco smoke: emerging evidence and arguments for a multidisciplinary research agenda. National Institute of Environmental Health Sciences NIEHS.

Gifford, D. (2017, March 8). 100 Things you can (and should) compost. Retrieved from http://www.smallfootprint-family.com.

Gruson, L. (1982). Color has a powerful effect on behavior, researchers assert. *The New York Times*, Science Section.

Hornikx, M. (2016). Ten questions concerning computational urban acoustics. *Building and Environment, 106,* 409–421.

Karagulian, F., Belis, C.A., Dora, C.F., Pruss-Ustun, A.M., Bonjour, S., Adair-Rohan, H., Amann, M. (2015). Contributions to cities' ambient particulate matter (PM): A systematic review of local source contributions at global level. *Atmospheric Environment, 120,* 475–483.

Kays, S., Posudin, Y. (2010). Volatile organic compounds in indoor air: Scientific, medical and instrumental aspects. National University of Life and Environmental Sciences of Ukraine.

Nicholas, S.W., Jean-Louis, B., Ortiz, B., Northridge, M., Shoemaker, K., Vaughan, R., ... Hutchinson, V. (2005). Addressing the childhood asthma crisis in Harlem: The Harlem Children's Zone Asthma Initiative. *American Journal of Public Health, 95*(2), 245–249.

Sullivan, W.C., Li, D. (2016). Impact of views to school landscape on recovery from stress and mental fatigue. *Landscape and Urban Planning, 148,* 149–158.

Veisten, K., et al. (2012). Valuation of greenwalls and green roofs as soundscape measures: Including monetised amenity values together with noise-attenuation values in a cost-benefit analysis of a greenwall affecting courtyards. *International Journal of Environmental Research and Public Health,* 3770–3788.

Wells, N., Evans, G. (2003). A room with a view helps rural children deal with life's stresses. *Environment and Behavior, 35*(3), 311–330.

Wolf, K.L., Flora, K. (2010). Mental health and function—A literature review. In: *Green Cities: Good Health*. College of the Environment, University of Washington. Retrieved from www.greenhealth.washington.edu.

Wolverton, B.C. (2012). *References and Technical Reports While Employed with NASA*. Retrieved from www.wolvertonenvironmental.com.

Wolverton, B.C., et al. (1984). Foliage plants for removing indoor air pollutants from energy-efficient homes. *Economic Botany, 38*(2), 224–228.

Wolverton, B.C., et al. (1989). *A Study of Interior Landscape Plants for Indoor Air Pollution Abatement: An Interim Report*. NASA.

Wolverton, B.C., Wolverton, J.D. (1993). Plants and soil microorganisms: removal of formaldehyde, xylene and ammonia from the indoor environment. *Journal of Mississippi Academy of Sciences,* 38(2), 11–15.

World Health Organization. (2010). *WHO Guidelines for Indoor Air Quality: Selected Pollutants*. WHO Regional Office for Europe.

World Health Organization. (2016). *Urban Green Spaces and Health*. WHO Regional Office for Europe.

Index